James Braid, Arthur E. Waite

Braid on Hypnotism. Neurypnology

Or, the rationale of nervous sleep considered in relation to animal magnetism or

mesmerism and illustrated by numerous cases of its successful application in the

relief and cure of disease

James Braid, Arthur E. Waite

Braid on Hypnotism. Neurypnology
Or, the rationale of nervous sleep considered in relation to animal magnetism or mesmerism and illustrated by numerous cases of its successful application in the relief and cure of disease

ISBN/EAN: 9783337240578

Printed in Europe, USA, Canada, Australia, Japan

Cover: Foto ©berggeist007 / pixelio.de

More available books at **www.hansebooks.com**

BRAID ON HYPNOTISM

BRAID ON HYPNOTISM

NEURYPNOLOGY

OR

THE RATIONALE OF NERVOUS SLEEP CONSIDERED IN RELATION
TO ANIMAL MAGNETISM OR MESMERISM AND ILLUSTRATED
BY NUMEROUS CASES OF ITS SUCCESSFUL APPLICATION
IN THE RELIEF AND CURE OF DISEASE

BY

JAMES BRAID, M.R.C.S., C.M.W.S., &c.

A NEW EDITION

EDITED WITH AN INTRODUCTION BIOGRAPHICAL AND BIBLIOGRAPHICAL
EMBODYING THE AUTHOR'S LATER VIEWS AND FURTHER
EVIDENCE ON THE SUBJECT

BY

ARTHUR EDWARD WAITE

LONDON

GEORGE REDWAY

1899

PREFACE

JAMES BRAID, the Manchester surgeon, is regarded as "the initiator of the scientific study of Animal Magnetism,"[1] and this is a title to distinction which possesses the guarantee of permanence. His method and the name which he applied to it are of world-wide use and knowledge, but the works in which he described and defended his discovery have never been reprinted, and are at the present day so rare as to be generally unknown. To collect these curious and, in some respects, highly important pamphlets for the purposes of the present edition has involved considerable research, as in some cases they are not in our national library.[2] This observation applies not only to " Satanic Agency and Mesmerism, a Letter addressed to a Clergyman,"

[1] Binet and Féré: *Le Magnétisme Animal,* c. 3. But this moderate statement by no means represents the whole sentiment which has found expression in France on the subject. Thus Durand de Gros, in the first hypnotic enthusiasm of 1860, declared that " the education and medicine of the soul found in Braidism a weapon of unheard-of power, which alone entitled the discovery of Braid to rank among the most glorious conquests of the human mind," *Cours de Braidisme,* p. 111.

[2] Moreover, no biographical notice of Braid, and no criticism, scientific or otherwise, is in agreement with any other in giving account of his published writings, nor is there one which can be said to be complete. The best bibliographical lists are to be found in the " Dictionary of National Biography," and in Preyer's German translation, but both are defective.

v

1842, which was Braid's first publication in connection with his discovery, but to " Electro-Biological Phenomena and the Physiology of Fascination," 1855, as well as several later works. Recourse in the first instance has been had to the Manchester papers of the period ; the second originally appeared in the *Monthly Journal of Medical Science*, and has been available in this form. For two later pamphlets, fortunately not of importance, a German translation has been alone available.

The chief treatise of Braid, containing the first formal and authoritative account of his experiments, is here reprinted without abridgement, and includes the original dedication, preface, and notes. It was published in 1843, under the title of " Neurypnology, or the Rationale of Nervous Sleep." Some time previously to 1846 the impression was exhausted, and a second revised edition was announced from time to time,[1] and as frequently postponed, till the sudden death of the author put an end to the project. The remaining writings of Braid, including several uncollected contributions to medical journals, have been subjected to careful analysis, are fully described in the introduction, and all passages of importance which are contained in them have been added in an appendix to illustrate the

---

[1] In the errata slip attached to " The Power of the Mind over the Body " (1846) ; in the preface to " Observations on Trance " (1850) ; and in the preface to " Observations on J. C. Colquhoun's History of Magic " (1852).

longer work. It is claimed, therefore, that this edition is representative and substantially complete. The omissions are of three kinds:—(a) Restatements of previous arguments, without important variations, and of experiments previously recorded in a more detailed manner ; (b) Observations on subjects extrinsic to the main issues of hypnotism, as, for example, the analogous effects of certain oriental drugs, and testimonies collected from sources outside the experience of Braid, as, for example, the interment of fakirs while in a state of trance; (c) certain controversial and minor matters, the interest of which has entirely passed away, e.g., the discussion with J. C. Colquhoun, author of " Isis Revelata " and other works, which arose largely through mutual misconception, and was set at rest before the appearance of the pamphlet which represents it.

Several translations of Braid's works are said to have appeared in French and German, but it is only necessary to notice the collected edition in the latter language which was published by W. Preyer in the year 1882.[1] It contains nearly all Braid's writings,

[1] *Der Hypnotismus.* Ausgewählte Schriften, von J. Braid. Deutsch herausgegeben von W. Preyer, Professor der Physiologie an der Universität, Jena. Berlin, 1882. It contains translations of several pamphlets which are entirely unknown in England, and great credit is due to the sagacity in research which has been rewarded by their discovery. The French translation published in 1883 is said to contain the posthumous work noticed in Appendix IV.

with the exception, curiously enough, of his main treatise. The substance of this had, however, been given in an earlier publication of the same writer,[1] who has also produced an original treatise on Hypnotism, in which a section is devoted to Braidism.[2] The industry of Preyer has indeed made the work of James Braid much better known in Germany than it is in this country, but it is hoped that the present edition will remedy this defect.

[1] *Die Entdeckung des Hypnotismus*, Berlin, 1881.

[2] *Der Hypnotismus.*   Vorlesungen gehalten an der K. Friedrich-Wilhelms, Universität zu Berlin.   Vienna and Leipsic, 1890.

CONTENTS

PAGE

PREFACE . . . . . .

BIOGRAPHICAL AND BIBLIOGRAPHICAL INTRODUCTION   I

HYPNOTISM OR NEURYPNOLOGY, THE RATIONALE OF NERVOUS SLEEP

ORIGINAL DEDICATION . . 69

AUTHOR'S PREFACE . . 71

INTRODUCTION

General remarks—Why Hypnotism has been separated from Animal Magnetism—How far considered useful in the cure of disease—Its powers on the animal functions—Certain erroneous charges refuted—Opinions and practice of Bertrand and Abbé Faria, Mr Brooks, Dr Prichard—Its moral influence —Should be used by professional men only—Definition of terms 83

PART I

CHAPTER I

Introductory remarks—Circumstances which directed to the investigation—A real phenomenon observed—Experiments instituted to prove the cause of it—Opinions and conclusions drawn from them—Reasons for separating Hypnotism from Mesmerism—Hypnotism more generally successful than Animal Magnetism—Mr Herbert Mayo's testimony on this point—Proofs of this referred to—Different modes by which it can be induced—Can only be expected to succeed by complying with the whole conditions required . . . 97

CHAPTER II

PAGE

Mode of Hypnotizing—Circumstances necessary to be complied with—Peculiar phenomena which follow ; excitement first, and afterwards depression of function—Importance of attending to this—How these may be made to alternate with each other—Extraordinary influence of a current of air during Hypnotism — Reasons for certain modifications of original modes of operating—Hypnotism proceeds from a law of the animal economy — Arises from the physical and psychical condition of the patient, and not from any emanation or principle proceeding from others—Example for proof—Exhibits no appreciable electric or magnetic change — Two patients may hypnotize each other by contact — Phenomena arise spontaneously in course of disease—Mr Wakley's admission on this point—Mr H. Mayo's testimony as to the effects of Hypnotism — Effects of different positions of the eyes—Remarks on articles in Medical Gazette—Consensual adjustment of eyes—Effects on size of pupil—Power of habit and imagination—Docility of patients, and exalted sensibility, and their effects—Patient hypnotized whilst operating on another—Mode of resisting influence   109

CHAPTER III

Phenomena of natural sleep, dreaming, and somnambulism contrasted—Causes of common sleep—Of dreaming—Effects of variety and monotony compared—Changes alleged to take place in the structure of the brain by exertion—Cause of Hypnotism, M'Nish's article on " Reverie" compared with Mr Braid's theory of Hypnotism—Mode of arousing patients from the state of Hypnotism   122

CHAPTER IV

Phenomena of common sleep—Of Hypnotism—Power of locomotion and accurate balancing of themselves—Tendency to dance on hearing appropriate music — Grace displayed under its influence—Tendency to become cataleptiformly fixed in any position if left quiet—Probability of Hypnotism having been practised amongst the ancients, and the cause of their superior excellence in sculpture, painting, and dancing—Effects analogous to Nitrous Oxide in some—in what it differs from this and intoxication from wine and spirits—Analogous to conium —Effects of monotonous impressions on any of the senses—Opinions of Cullen, M'Nish, Willich—(Counting and repeating generally known)—A writer in *Medical Gazette*—Power of

PAGE

habit and expectation—All the phenomena consecutive—Note illustrating this—Power of Hypnotism to cure intractable diseases and disorders—Miss Collins's case and Miss E. Atkinson's—Extent to which it may be expected to be useful . 130

CHAPTER V

Reasons for delivering Public Lectures on Hypnotism—Mode of procuring refreshing sleep, with low pulse and general flaccidity of muscle—Efficacy of this plan 146

CHAPTER VI

Introductory remarks—Relation between mind and matter illustrated to disprove materialism—Armstrong, Colton, Brown, Abercrombie, Stewart, Plato—General conclusion, that mind or life is the cause of organism—Power of conscience—The passions, how excited—Dr Elliotson's opinion as to the efficacy or non-efficacy of volition and sympathy with operator's brain —Modes of dividing the brain—Causes of Phrenology being imperfect—Objections to Phreno-Hypnotism—Mode of connection between the brain and body—First attempts at Phrenologizing during Hypnotism were failures—Succeeded by operating differently—Cases illustrative—A child operates successfully—Details of the case—Other successful cases— Proper time for operating—Case of an officer in 1758—Inferences as to its curative powers—Opinions of La Roy Sunderland, and Mr Hall—Presumed cause of phenomena called Cross Magnetism—Return stolen property to proper owners and proper place by smell and touch—Power of hearing *faint sounds*—Additional cases—Opposite faculties can be excited at same time by acting on the opposite hemispheres—Mode of operating—Concluding remarks on the value of testimony . 150

CHAPTER VII

General resumé—Many phenomena admit of physical and chemical proof—Difficulties of comprehending many phenomena— Effects of prejudice in preventing the reception of truth— Critique on debate at Medico-Chirurgical Society on Mr Ward's operation—State of the circulation—Conjectures as to the cause of the cataleptiform condition . . . 216

PART II

PAGE

Modes of operating—Objects of operations—Cases of sight improved—Hearing—Deaf and dumb—Mr Curtis's remarks—James Shelmerdine's case — Mr Bingham's testimony — S. Taylor—Sense of smell—Touch and resistance—Tic, paralysis of sense and motion, cured—Miss E. Atkinson's case, voice recovered—Rheumatism, ten cases—Irregular muscular action —Headache—Spinal Irritation—Epilepsy—Spinal curvature —Neuralgia, and palpitation of the heart—Surgical operations without pain—Diseases of the skin—Locked jaw—Tonic spasm —Miss Collins—Concluding remarks 225

APPENDIX I

NOTES BY THE EDITOR, CHIEFLY DERIVED FROM THE LATER WRITINGS OF JAMES BRAID   . 321

APPENDIX II

NOTES OF FURTHER CURES BY HYPNOTISM FROM THE LATER WRITINGS OF JAMES BRAID   . 340

APPENDIX III

SYNOPSIS OF COUNTER-EXPERIMENTS UNDERTAKEN BY JAMES BRAID TO ILLUSTRATE HIS CRITICISM OF REICHENBACH   352

APPENDIX IV

A POSTHUMOUS WORK OF JAMES BRAID   .   . 362

APPENDIX V

A SHORT BIBLIOGRAPHY OF THE WRITINGS OF JAMES BRAID   .   .   . 364

INTRODUCTION

§ 1. THE LIFE OF JAMES BRAID.

THE materials for the biography of James Braid are
so meagre that a paragraph of modest dimensions
will be found to exhaust them ; nor is the reason far
to seek. His life would seem to have been of an un-
chequered and even character ; it was free from early
struggles, for it was entered under favourable circumstances, and it was moderately successful throughout,
but with no approach to brilliance, being lifted out of
the groove of the ordinary competent practitioner in
the north of England by a single great event ; by this
event also it was ruled, and all details of a domestic
and private kind disappear beneath the shadow cast by
the discovery of Hypnotism. Braid was born on the
estate of his father, a landed proprietor of Fifeshire,
about the year 1795 ; he was educated at the University of Edinburgh, and, as appears by the dedication
of his chief work, he was apprenticed to Dr Anderson
of Leith and to his son Dr Charles Anderson. Having
obtained his diploma of M.R.C.S.E., he received an
appointment as surgeon to the miners on the Hopetoun
works in Lanarkshire, but afterwards removed to Dum-
fries, where he practised with Dr Maxwell. Here a stage-
coach accident brought him a patient who was a visitor
from a distance. Considerably injured, Mr Petty was

much impressed by the talents and attention of the young surgeon, whom he persuaded to remove again ; and thus it was that James Braid came to be associated with Manchester. There he made his discovery, practised his cures by means of it, wrote his pamphlets, earned himself position and respect by his general skill, and popularity by his pleasant disposition. There he married, and there also he died on March 25, 1860. His end was most unexpected ; he went to rest on the previous night in his usual good health, but the next morning awoke with an acute pain in his side ; after drinking a cup of tea, he breathed heavily twice or thrice, and expired. The cause of death, though probably still ascertainable by reference to the registers of the period, has never been specified distinctly, but it was attributed by supposition to heart disease.[1]

§ 2. Epochs in Animal Magnetism prior to Braid's Discovery.

Shortly before the birth of James Braid the first epoch in the history of Animal Magnetism may be said to have closed. The government commission nominated in France to report on the alleged fluid of Mesmer, had announced its decision, and another, composed of the members of the Royal Society of Medicine, had followed almost immediately. Their hostile verdicts were not appreciably affected by the minority report of Laurent de Jussieu, member of the second commission, who, believing that the *illuminé* of Vienna was on the track of a pregnant

---

[1] *Lancet*, March 31, 1860.

truth,[1] had the courage to maintain this view. Prometheus or impostor, Mesmer's career was at an end ; the day of trough and baton, of the Hotel Bullion and the coat of lilac silk, was succeeded by a day of obscurity, and of such exile as foiled ambition obtains in its native country. He returned to Germany and survived, almost forgotten, until the epoch of Waterloo. His death was, broadly speaking, coincident with Braid's reception of a medical diploma.

The second period in the history of Animal Magnetism began when the Marquis de Puységur discovered artificial somnambulism, and Pététin, a physician of Lyons, made the first experiments in cataleptic transposition of the senses.[2] It may be said to have ended, shortly before the death of Mesmer, with the publication of the celebrated " Critical History of Animal Magnetism " by the naturalist Deleuze.

In the year 1825 the subject entered apparently upon a more favourable period, when the French Academy of Medicine, the successor of the Royal Society, nominated a third commission of inquiry, which lasted for five years, and issued a favourable report in the year 1831. The Academy declined to print it, and a fourth commission published in 1837 a report which was decidedly hostile. At this time Braid was settled in Manchester and was, quite unconsciously, on the eve of his discovery.

What was the position of the subject in this country

---

[1] He based his opinion on what he termed a number of well-established facts, independent of imagination, and for him beyond doubt. By these he was led to admit the existence or possibility of an agent transmissible from one human being to another, sometimes by a simple approximation. He sought to identify this agent with animal heat.

[2] His experiences are recorded in a work entitled " Animal Electricity."

at the period in question? Writing in the year 1839, the Rev. Chauncy Hare Townshend observed that the majority of Englishmen have derived their ideas of mesmerism from the experiments of Dr John Elliotson, and there can be no doubt that the English history of Animal Magnetism centred at the period in this eccentric personality, and he remains its typical representative, as he was indeed its most active exponent. A man whose untiring energy had largely assisted the foundation of University College Hospital, a professor of Practical Medicine in the University of London, and the President of the Royal Medical and Chirurgical Society, it is difficult to understand how his testimony could be set aside by his professional brethren, but the fact itself is indubitable, and it is not less true that Dr Elliotson suffered irretrievably for espousing the detested cause. It is partially unaccountable at the present day that feeling should have been so strong in the matter ; the physiological conditions induced, or supposed to be induced, by Animal Magnetism did not so seriously threaten the established positions, or even the accepted preconceptions, of medical science as was the case later on with Spiritualism when the exponents of general science had pronounced definitely for materialism, and the allegation of " spirit return " constituted a direct challenge to the message of modern science. But the hostility was honest in its way, though based on *à priori* considerations which are somewhat humiliating to recall, as, for example, that mesmerism had been practised by impostors and quacks ; that most of its believers were ignorant or mere smatterers ; and that many of their stories were undeniably exaggerated, if not actually invented. These were good grounds for caution, but they were no grounds at all for passing over the evidence of those who were not charlatans, whose integrity and ability had never been questioned previously, whose statements were strongly documented. However, they were the rule of the moment, they justified a persecution which could not well have been more bitter if Elliotson had been a convicted cheat. They forced him to resign his professorship, they destroyed his practice, and there was a perceptible tone of triumph when his ruined prospects were recited.

At the same time there were persons committed to neither issue who saw that the contempt and incredulity excited by the conversion of Elliotson were largely rooted in an illiberal prejudice, and these were not backward in demanding a dispassionate investigation. The foundation of the *Zoist*, a quarterly magazine containing the most elaborate records of experiments in the curative powers of Animal Magnetism, gave additional force to the demand, and in the year 1843 there is ground for believing that the hostile parties were conscious of their strained position. In that year James Braid came forward as something little short of a saviour of society, and he made retreat honourable. The publication of his work on " Neurypnology, or the Rationale of Nervous Sleep,"[1] created a new era in the history of Animal Magnetism, and was destined to change in a material manner the relations previously subsisting between the two parties.

[1] The first contribution which Braid offered to medical literature was " Observations on Talipes, Strabismus, and Spinal Contortion," published in the *Edinburgh Medical and Surgical Journal*, 1841.

§ 3. THE INCEPTION OF HYPNOTISM.

There is no need to review at any length the account which Braid has given of his discovery, as it is the design of this edition to place it once more before the public, in accordance with the expressed intention of the author himself, but it will be useful to consider shortly the effect of that discovery before it was officially announced by the appearance of the work in question. Had the travelling French mesmerist, Lafontaine, never visited Manchester, the American Grimes might have been in a position to claim something more than an independent origin for his so-called Electro-Biology, or, what is more probable, Hypnotism, in any form, might not have been heard of for many years to come. The fortunate accident of this visit seems to call for some further commemoration than Braid has given it, and, as it so happens, the materials are not wanting.

In the year 1866, M. Ch. Lafontaine published his "Memoirs of a Magnetiser"[1] in two stout volumes, conceived in the heroic spirit and elaborated with due regard to the momentous nature of the character in chief. It is instructive to compare his account of his séances at Manchester with the brief and sober narrative of James Braid. As it concerns the inception of Hypnotism, it may be permissible to reproduce it at some length; it will also afford an opportunity to judge how memoirs are occasionally manufactured in France, more especially in matters of the marvellous.

"Having accomplished the cure of numerous deaf and blind persons," says M. Lafontaine,[2] with modest

[1] *Mémoires d'un Magnétiseur, par Ch. Lafontaine, suivis de l'examen phrénologique de l'auteur par le Docteur Castle.* 2 vols. Paris, 1866.
[2] *Ib.*, vol. I., c. 11.

assurance, " as also of numerous epileptic and paralytic
sufferers, at the Hospital "—this was in Birmingham—
" I repaired to Liverpool, but only to meet with dis-
appointment; few persons attended the séance ; and on
the following day I proceeded to Manchester, one of
those cities in which my success was conspicuous. The
newspapers reported my experiments at great length,
filling four or five columns of the colossal English
journals, and to give some idea of the sensation which I
created in this great manufacturing centre, I may say
that my séances returned me a gross total of thirty
thousand francs, though the charge for admission was
only half-a-crown. The concourse was truly immense,
and the lecture hall of the Athenæum was more than
crowded. I put to sleep a number of persons who were
well-known residents of Manchester, occupying good
positions, among others Messrs Lynnil, Higgins, Dyren-
furth, a gentleman on the staff of the *Guardian*, and
several more. I caused deaf mutes to hear, operated
a number of different brilliant cures, and then retired
to Birmingham a second time, where I had an engage-
ment to visit a patient, and was to give some further
demonstrations. After my departure, Dr Braid, a
surgeon in Manchester, delivered a lecture in which
he proposed to prove that magnetism was non-existent.
From this lecture Braidism, afterwards called Hypnotism, originated, ardent discussions arising, even from
the beginning, over this pretended discovery. I received letters entreating me to return to Manchester,
and I did so on a date when Dr Braid had announced a
demonstration at the Mechanics' Institute. I attended
with two other persons, taking seats in the gallery, so as
to see without being seen ; but I was recognised by the

managers, who begged me to come down and occupy a chair on the platform. This I should personally have preferred to decline, for although Dr Braid had shown me some opposition, I did not wish to exhibit any in return, but simply to devote myself to private obser- vations. However, I could not well refuse the request of the managers who pressed the point in exceedingly flattering terms. When the door at the foot of the platform opened, and I appeared, there was deafening applause, and cries of 'Bravo, Lafontaine!' etc. I bowed ; the applause, the cries, the shouts were re- doubled. I again bowed, and was about to sit down, but still the salvo continued with increased enthusiasm. I advanced thereupon to the edge of the platform, and as I did so the cheering of three thousand spectators became more overwhelming than ever. At length I prayed for silence, and continued doing so, using ex- pressive gestures, but nothing could stay the outburst ; it was an ovation, a veritable triumph. At last a voice shouted : 'We want M. Lafontaine's opinion on Dr Braid's experiments.'

" ' Gentlemen,' I replied, ' I have come from Birming- ham expressly to watch the demonstrations of Dr Braid, and to accept with enthusiasm whatever may be true in regard to them.'

" At these words the applause was renewed more vociferously than ever, whilst I returned to my seat. There was a pause, however, upon the entrance of Dr Braid, but it broke out a third time when he and I shook hands. His experiments began, but, unfortun- ately, on this occasion none of them succeeded ; neither sleep nor catalepsy was obtained, and every moment I was appealed to, with ' What is your opinion, M. Lafontaine?' I did not reply, but towards the close of the demonstrations, the appeals became so universal and so urgent that I could not escape saying a few words.

" ' Gentlemen,' I observed at length, 'I engage to give my views upon what has just taken place at a special séance when I will myself perform Dr Braid's experiments, and will also explain them.'

" The applause was renewed and the meeting terminated in failure. In the facts that were advanced on this occasion by Dr Braid, there was in my opinion absolutely nothing that was remarkable, and had not that gentleman been honourably known in the town, I should have supposed that he was mystifying his audience. The next day, and for six days consecutively, I experimented after his own fashion on fifty or sixty subjects, and the results were practically *nil.* I then gave a magnetic séance ; I placed the subjects of the disc[1] side by side with magnetic subjects; the experiments made upon the first proved once more abortive, while those made upon Eugène and Mary[2] were invariably marked and positive in their results. Later on, Dr Braid and others joined magnetism to the pretended science at that time termed Braidism, and it should be noted that when members of the faculty finally introduced hypnotism to Paris, they only obtained genuine and precise results after making passes over their subjects."

All things considered, the most noteworthy feature of this amazing narrative is not the emblazonment which it

[1] At the inception of his experiments Braid made no use of a disc, which indeed was first introduced by the American school of electro-biologists.

[2] The habitual subjects of Lafontaine, whom he carried from town to town, and on whom he could rely for phenomena.

has received at the hands of the artist, but that it should
possess so much foundation as it proves on examination
to have. Before M. Lafontaine visited Manchester, the
report of his séances in the metropolis had reached the
north, and had been a subject of highly unflattering
comment at the hands of the *Manchester Guardian;*
there was therefore no preconception in his favour, and
the first of his series of demonstrations was only curtly
noticed in that newspaper. But his original audience
of perhaps two hundred persons was increased on subsequent occasions till the hall was filled to overflowing,
and though the profits of the series are placed at an
impossible figure, they were undoubtedly handsome.
The *Manchester Guardian* subsequently gave reports of
the proceedings which were in proportion to the interest
excited, and the dimensions mentioned by Lafontaine
were actually reached. Braid was present at the second
and nearly all further demonstrations, and it was the
active, though hostile, part which he took, in conjunction with other medical men, that animated the proceedings. The sympathies of the audience, at least in
the first instance, were largely with the mesmerist, and
possibly Braid did not show the best taste at the
beginning. He was more than repaid for any initial
rebuffs, however, as he more than atoned for any initial
mistake, when he undertook his counter experiments.
He also had the satisfaction of attracting crowds to his
lectures, though the profits were not claimed for himself,
and he was reported at no less length.

The second visit of Lafontaine did actually take place,
and he was present at one of Braid's demonstrations,
but he experienced no such reception as he describes,
and the active part which he took was confined to feel-

ing the pulse of one or two of the subjects. Lafontaine, it should be added, was not acquainted with English and communicated with his audience through an interpreter. Finally, Braid's experiments were on this occasion quite as successful as previously. For the rest, Lafontaine's pretended attempts to reproduce hypnotic phenomena were mere travesties, and were denounced as such by his opponent. Like other mesmeric lecturers the Swiss magnetiser brought his subjects with him, and his success with strangers was comparatively poor. Read over at this date, the controversy of the period seems to show that Braid was mistaken in seeking publicity at so early a stage of his discovery; his demonstrations and his conclusions alike suffered in consequence. But for this also his subsequent caution and moderation made full amends, though, on the other hand, his haste gave impetus as it gave notoriety to his views, and both might have been lost by delay.

At the close of the year 1841, a London lecturer, named Duncan, took up the subject of Braid's discovery, reproduced his experiments, and announced the approaching visit of the Manchester surgeon, who began already to be regarded as a person of some eminence. Correspondents of the *Medical Times* deposed that they had had recourse to the mode of practice adopted by the new theory, and had found it successful; the lock of Animal Magnetism was said to have been picked, and in recognition of the time and money sacrificed by Braid over his discovery, it was even suggested that a suitable acknowledgment should be made to him " in the shape of a piece of plate."

At the beginning of March 1842, Braid arrived in London, and gave lectures at the Hanover Square

Rooms and at the London Tavern. These were well
reported and seem to have been successful in all respects.
While on this visit, he contributed a letter to the *Medical
Times*, in which he stated that he had operated on the
blind, and had thrown them into the hypnotic state,
from which he concluded that it was the ganglionic or
sympathetic systems and motor nerves of the eye, as
also the state of the mind, which were affected by the
process, rather than the optic nerve itself.[1] On his
return journey he gave lectures in Birmingham, and
subsequently another course in his own city. In April
1842, Lafontaine paid a third visit to Manchester, and
the rival methods became again the subject of ardent
debate. In the midst of this debate, the Rev. James
M'Neil, apparently a clergyman of the district, undertook to exhibit the diabolical character of all mesmeric
operations, which at that period was not an unfamiliar
accusation, and Braid replied in a pamphlet, which,
though it is not of any special moment as regards his
discovery, is practically the first memorial concerning
it, and is therefore of some interest.

§ 4. Reception of the Discovery.

Though Braid survived his discovery by not more than
eighteen years, he lived to know that it was well on
the road to acceptance by the competent opinion of the
time.[2] It possessed the recommendation which belongs

[1] Mr Herbert Mayo, a London member of the profession, was about this
time one of Braid's witnesses, and recorded his unqualified concurrence
in the general conclusions.

[2] "I feel no great anxiety for the fate of *Hypnotism*, provided it only
has 'a fair field and no favour.' I am content to bide my time, in the firm
conviction that truth, for which alone I most earnestly strive, with the
discovery of the safest, and surest, and speediest modes of relieving human
suffering, will ultimately triumph over error."—"Magic, Witchcraft," etc.,
p. 53.

to the middle way. The practice of Animal Magnetism had contrived to survive opposition of a peculiarly violent character ; its importance was increasing ; its supporters at the moment included men who were representative in their several spheres ; the adherence of an independent and accomplished mind like that of Archbishop Whately must have exerted considerable influence ; cases of this kind made it less and less possible to over-ride the alleged facts on grounds of an arbitrary nature. Its professional supporters, moreover, were not persons who could be silenced by proscription ; they were in many cases men of accepted scientific repute, whose positions would have been above challenge if they had not espoused mesmerism. Now, it was the obnoxious hypothesis of the magnetic fluid rather than the well-attested percentage of sober facts which created the hostility,[1] and a record of experiments which fully established the chief facts but exploded the previous explanation was honestly welcome, if only because it relieved the tension of the existing position. It put also the hostile verdicts of the French commissions of inquiry in a new light. Those labours had

[1] " Even the French commission of 1784 bore testimony *to the reality of the phenomena*. What they disputed was Mesmer's power of proving to them the existence of a magnetic fluid as alleged by Mesmer. This was the real extent of their condemnation, which has been so constantly and triumphantly referred to in opposition to the pretensions of mesmerism."— From a letter on Mesmerism addressed by Braid to the *Medical Times*, under date of December 16, 1843. In France, Count Agénor de Gasparin, a contemporary of Braid, gave expression, a few years later, to much the same view, observing that the Commission was careful not to proceed too far in its denials, and adding further that the method which it adopted laid it open to criticism. (*Des Tables Tournantes*, vol. ii. pp. 281, 282.) This writer's moderate and discriminating views of Animal Magnetism make one regret that he does not appear to have been acquainted with the experiments and conclusions of Braid.

been directed chiefly to the evidence for the alleged
magnetic fluid ; the reports condemned the hypothesis
without giving anything in the shape of a final and
unanswerable verdict as to the occurrences.[1] Furthermore, Braid's researches had wholly failed to establish
the higher phenomena of thought-reading, clairvoyance and the ability of the somnambulists to diagnose
diseases. Many persons in the scientific world therefore lent a willing ear to the demonstrations of hypnotism, though in the circles which dispense reputation to
medical specialists and give currency to new medical
discoveries, it must be confessed that there was a certain
hostility at the inception. The *Lancet* at that period
had pronounced definitely and finally against mesmerism,[2] and books in support of this subject were
very seldom reviewed in its columns, unless by their
weakness or extravagance they offered a special opportunity to hostile criticism. Braid's publications passed
entirely unheeded,[3] so far as any formal notice was concerned, from the date of " Neurypnology " to the date
of the author's death. The theory of hypnotism and
the experiments by which it was substantiated were

---

[1] The criticism of Arago may be quoted in this connection : " Effects
analogous or inverse might evidently be occasioned by a subtle, invisible,
imponderable fluid, by a species of nervous, or if preferred, a magnetic
fluid, circulating in our organs. Hence the commissioners abstained from
talking of *impossibility* in connection with this subject. Their thesis was
more modest, they were content with laying down that there was *nothing
to demonstrate* the existence of such a fluid."—*Annuaire de* 1853, p. 437.

[2] " Mesmerism is too gross a humbug to admit of any further serious
notice. We regard its abettors as quacks and impostors. They ought to
be hooted out of professional society."—*Lancet*, Oct. 29, 1842. Later on
(Feb. 4, 1843), the same organ admitted that the subject offered a mass of
material to the study of psychic medicine—about which discrepancy the
Lancet's rancorous rival observed, *Quem Deus vult perdere dementit.*

[3] See, however, Appendix V., No. 21.

mentioned indeed by Dr Radclyffe Hall in the crude and ill-digested account of the "Rise, Progress, and Mysteries of Mesmerism," which ran as a serial in the journal. But Braid had occasion to complain of the presentation thus made of his opinions, "coming as it did immediately after a collocation of all the extreme views of the *ultra mesmerists* . . . without a single qualifying remark." Dr Hall's tabulation does not, however, seem to have been intentionally unfair,[1] and he went so far as to explain in a footnote that Braid was "by no means to be enrolled among the supporters of mesmerism, as commonly understood, many of his views being ingenious and philosophical." But he had evidently been at no great pains to possess himself of the case for hypnotism, or he would scarcely have defined it as differing from mesmerism merely by the fact that it did not require "the assistance of a second person to produce the effects."

When the work on Neurypnology appeared the *Lancet* had a rival in the shape of another weekly periodical, namely, the *Medical Times*, then recently established, and there was a bitter feud between them. The fact that mesmerism in all its forms was detested by the one perhaps made it less unpalatable to the other,[2] and although it remained non-committal, it opened its

[1] Indeed his general conclusion approximated to that of the discoverer of hypnotism, being, as he himself described it later on, "that all the ascertained and probable effects of mesmeric procedure depended on a peculiar state of the nervous system, which might be induced in a variety of ways, and not upon any imagined fluid given (*sic*) by the operator to his patient." (*Lancet*, April 24, 1847.)

[2] A leading article under date of July 22, 1843, seems to adopt a very reasonable standpoint, observing that whatever may be the claims of mesmerism on our respect, "there is no disputing longer the force of its demands on our attention." Again, "we *are* not, and, as philosophers, *cannot* be justified in treating it with *unenquiring* incredulity."

columns to the discoverer of hypnotism, who was a frequent contributor during the few years that remained to him, and seems to have interested its readers. The subject thus obtained advertisement, the work which announced it sold rapidly, creating much inquiry; and as years went on, Braid was able to cite approved scientific authorities in favour of his discovery, and was justified in saying: "All I claim for hypnotism is now willingly admitted by the great majority of scientific men, who have investigated the subject without previous prejudice in favour of mesmerism." Professor Alison and John Hughes Bennett were among his supporters. Dr Carpenter regarded his experimental researches as throwing "more light than has been derived from any other source upon the phenomena of mesmerism."[1] Garth Williamson described them as the invention of a man of genius, and a few liberal exponents of mesmeric practice also confessed their value.[2] But in mesmeric circles generally the reception of hypnotism offered another illustration of the readiness with which those who have suffered persecution for justice' sake become persecutors to injustice in their turn. "The mesmerists," says Braid, "thinking their craft was in danger—that their mystical idol was threatened to be shorn of some

[1] Dr Holland, the Queen's physician, Sir David Brewster, and Dugald Stewart, the philosopher, should also be included. Durand de Gros says, that "the most distinguished scientists of England studied carefully the experiences of Braid and the electro-biologists; they bore witness to the reality of the phenomena, and loudly and conscientiously proclaimed its importance"—a view which slightly magnifies but does not distort the actual position.

[2] Professor Gregory and Dr Esdaile must, to their honour, be included among these. The former had been an eye-witness of Braid's experiments, and, in his "Letters to a Candid Inquirer," he observes: "I most willingly bear testimony to the accuracy of his description, and to the very striking results which he produces."

of its glory by the advent of a new rival—buckled on their armour, and soon proved that the *odium mes-mericum* was as inveterate as the *odium theologicum*; for they seemed even more exasperated against me for my slight divergence from their dogmas, than would have been the case had I continued an entire sceptic, denying the whole as a mere juggle or imposture. I could furnish ample proofs of the persecution and misrepresentation to which I have been subjected by the mesmerists, but the following example shall suffice. In consequence of my *hypnotic heresy*, and my honest endeavours to protect myself and hypnotism against the most unfair and wilful misrepresentation by a chief in the mesmeric school, this gentleman, who, as well as his friends, had raised a mighty outcry against the cruelty and injustice, illiberality and persecution which the medical profession had manifested towards *him*, in consequence of his having adopted the mesmeric notions and practice in some cases, carried his spleen and persecuting spirit against me to such a pitch, that he raised such determined opposition at headquarters as deprived me of an official appointment, which had been most kindly and voluntarily offered to be secured for me by the chairman of a public board; and which election, I have good reason to believe, would have been decided in my favour, but for the implacable opposition to me, for the above-named cause, of this illiberal, vindictive, and persecuting mesmeric autocrat." It is impossible to suppose that a man of Braid's known amiability could have spoken thus vehemently in the absence of full warrant, nor is it possible to mistake the person to whom the reference is made, and of whom enough is otherwise known to render it not antecedently

improbable. The first resource of the mesmerists was
of course the device of silence. As long as possible,
Hypnotism was a tabooed subject. The *Zoist*, under the
direction of Elliotson, ignored Braid for years. In other
quarters the discovery was first denied and then belittled
by the attempt to show that it was not new.[1] There is
evidence that this charge was felt somewhat keenly at
the outset, but as there was an element of truth at
the bottom of it,[2] Braid was not long in accepting the
inevitable with resignation, if not with relish. He
observed that his critics were as well aware as himself
that his method for demonstrating the fallacy of the

[1] Another objection, and one which appears characteristic, was that
Braid was an unconscious mesmerist, possessing an unusually powerful
magnetic temperament, to which his hypnotic successes were mainly
attributable. This was the opinion of Elliotson, for it is doubtless he to
whom Braid alludes when describing (" Magic, Witchcraft," etc., pp. 36,
37) his visit to a " chief in the mesmeric ranks," who, having read the
results of his experiments, " expressed himself sure that I had a large
brain, a large, capacious chest, and great mental energy, *i.e.*, that I possessed a determined will. He farther added, as a proof of his sagacity,—
' And now that I see you, you are just the person I supposed, for you have
them all.'" This point of view is not yet dead. Professor Coates, in his
" Human Magnetism," recently issued (Redway), maintains that Braid was
mistaken in ignoring his own personal influence and presence. " He
omitted to notice that in carrying out his experiments, he was in his physical and psychical prime. The records of his experiments seem further to
indicate either that he was peculiarly fortunate in obtaining suitable subjects or that he lived at a time when people were more susceptible than at
this day. It will be remembered that one of his sensitives, an illiterate
warehouse girl, was entranced by Braid in the presence of Jenny Lind, on
September 3, 1847, when she followed the Swedish Nightingale's songs in
different languages both instantaneously and correctly, and when, in order
to test her powers—which were natural but untutored—Mdlle. Lind ex-
temporised a long and elaborate chromatic exercise, she imitated this with
no less precision, though unable in her waking state even to attempt anything of the sort." (Carpenter's " Mesmerism and Spiritualism.")
[2] " Although suggestion was not new, Braid at least established and
demonstrated it."—Cressfield's " Value of Hypnotism," 1893. Though
little more than the circular of a professional masseur, Mr Cressfield's
pamphlet has several good points, well and moderately expressed.

mesmeric theory concerning a magnetic fluid, was entirely new, in so far as he was aware at the time, or was, in other words, regarded as a discovery by himself.[1] When he came, however, to investigate certain practices of the " Hindoos and Magi of Persia," he was made acquainted with " many statements corroborative of the fact, that the eastern saints are all self-hypnotizers— adopting means essentially but not identically . . . the same as those which I had recommended for similar purposes." Again, he cites as a curious and important fact, clearly demonstrated by his investigation, " that what observation and experience had led me to adopt and recommend as the most speedy and effectual mode of inducing the nervous sleep and its subsequent pheno- mena, had been practised by the Magi of Persia for ages before the Christian era—most probably from the earliest times."

.§ 5. LITERARY ACTIVITY OF JAMES BRAID.
FURTHER EXPERIMENTS IN PHRENO-MAGNETISM.

The publication of " Neurypnology " was followed immediately by two papers contributed to the *Medical Times*, under the title of " Observations on the Pheno- mena of Phreno-Mesmerism," and designed in fulfilment of a promise made in the preface to his first work, that his experiments on this subject should be extended. They constitute a further attempt to explain the power

[1] " Braidism possesses the fatal power of every great innovation which once having routed its enemies, completes their humiliation by compelling them, under I know not what mysterious attraction, to bury themselves in the dust of the libraries and to exhume with feverish zeal its titles to nobility, demonstrating by every kind of device that what they denied only yesterday as impossible and absurd existed in all times."—*Cours de Braidisme*, p. 23.

acquired during " mesmeric, hypnotic, or nervous sleep, of exciting the passions, emotions, and mental manifestations, through impressions made on different parts of the body,"[1] by seeking to reduce it to the laws of sympathy and association, connected with automatic muscular action; that is, " by titillating certain combinations of nerves, corresponding muscles are called into action, and this muscular action renovates the past feelings with which they are ordinarily associated during the waking condition, in the active manifestation of the various emotions and propensities of the mind." Braid in reality had not as yet added to his experiments, or at least did not publish their results, as he proposed ; what he really gave in these papers was a more matured statement of his previous conclusions with a few fresh points.

The alleged phenomena of Phreno-Mesmerism are sufficiently explained by the section devoted to this subject in " Neurypnology," which also establishes that they may be excited through the muscles of the trunk and extremities as well as through those of the head and face. The general result of Braid's experiments was to classify and curtail the phrenological organs rather than to endorse their fantastic multiplication, which was a craze of phrenologists at the moment. Nor did they do anything to prove or to disprove the central doctrine of the science, namely, the " allo-

---

[1] In the year 1842, two phrenologists, acting independently of one another, Dr Buchanan, of Louisville, U.S.A., and Mr J. B. W. S. Gardner, of Roche Court, Hants, claimed ability to excite or suspend the action of any cerebral organ by means of Animal Magnetism directed to that part of the head where such organ is affirmed to be situated. See *Phrenological Journal*, vol. xv., p. 188. Also Dr Engleduc's Introductory Address at the fifth Annual Session of the Phrenological Society, held in London, June 1842.

cation of different functions to separate organs in the brain." The most that was exhibited in this respect was the existence of sympathetic groups in the head, the lateral portions representing the animal propensities—"the prehensile or selfish principle"—while the central and coronal region, excepting the vertex, corresponded to the opposite mental condition, representing the moral faculties—"the relaxing, extensile, and distributive or benevolent tendency, respect and regard to others." The intermediate space seemed to rouse a mixed or varied character of the emotions under operation, while the forehead excited attention, observation and reflection.

It is noted by Braid as one of the most remarkable circumstances in connection with these investigations that "the influence of contact in exciting memory" is of a marked character in hypnotism, as it is also no doubt in its own degree during the ordinary waking state. "A patient," he observes, "may be unable to answer the most simple question on a subject with which he is quite familiar; but touch any part of his person, either with a finger or any inanimate substance, or cause him to place one of his own hands or fingers in contact with some other part of his own body and immediately he will be able to reply correctly." The explanation proposed was that the attention was thus fixed and concentrated in one direction instead of wandering uncontrolled, "whilst addressed only through the organ of hearing."[1]

---

[1] Braid afterwards observed that this influence of contact in exciting memory threw much light on the alleged *rapport* of the mesmerists, "as it proves the attention may be thus excited to receive impressions to which it would otherwise be quite dormant."

The second paper, while purporting to be in continuation of the same subject, abandons Phreno-Magnetism entirely, and is concerned chiefly with sanative
hypnotism, enumerating experiments by the writer,
some of which were subsequently republished elsewhere, and will be found in the second appendix to
this volume. Here it is only necessary to notice the
curious opinion, based on observation, that the hypnotic subject is frequently guided by an exalted sense
of smell, but describes that guidance in the terms of
the sense of sight.[1] "A patient securely blindfolded,
if asked to find out any one he knows in a room full
of company will readily do so by smell. He will tell
you he sees the person, but the moment the nose is
held he no longer sees him, and will turn the head
as if looking for the party; but the moment the nose
is unstopped he thinks he again sees him. In like
manner a glove or pocket-handkerchief being delivered
to a patient, without any possibility of knowing to
whom it belonged, if asked to deliver it to the proper
owner, he will readily find the proper party by smell.
I have thus seen a patient restore four white cambric
handkerchiefs to their proper owners, although huddled
together and put into the patient's hands at once, whilst
securely blindfolded. There was positive proof that
this was done by smell, as it was always determined
by smelling to the persons and handkerchiefs before
delivering them to the respective parties."[2]

[1] Compare "Neurypnology," Part I., c. 6, p. 193.
[2] By the same exaltation of olfactory sense, Braid afterwards explained
the efficacy of "mesmerised water" in a manner perhaps more curious
than convincing. The efficacy in question can be, in most cases, fully
accounted for by suggestion, as Braid indeed perceived, but the observations are worth noting, if only because they show that the alleged dis-

In the preface to "Neurypnology" there is a brief reference to the now well-established fact that "whatever images or mental emotions or thoughts have been excited in the mind during nervous sleep, are generally liable to recur, or be renovated and manifested when the patient is again placed under similar circumstances." In some "Observations on Mesmeric and Hypnotic Phenomena,"[1] Braid regards this fact, as evidence of double consciousness, and cites in the same connection an interesting experiment of his own.

"By double consciousness I mean that there is a stage of the sleep, when a patient may be taught anything, and be able to repeat it with verbal accuracy as often as in that stage again, whilst he may have no idea either of the subject or the words when in the waking condition. This is quite analogous to what is recorded of natural somnambulists; but it was some time before I discovered the exact stage for eliciting it, and I had consequently recorded this as a remarkable difference between natural and artificially induced

coverer of suggestion did not always apply his own theory except as a last resource. "There is an aroma peculiar to every individual, both as regards the exhalation from the lungs and skin; hence the dog can trace his master through a crowd by smell, or any one else he has been set on the trail of. Either breathing on water, or wafting over it with the fingers, therefore, might readily impart a physical quality to suggest, by means of smell, ideas connected with some individuals . . ., and the ideas thus excited, through the laws of association, would produce the wonted results. Nay, even the very *mode* of *presenting* it might be sufficient, of itself alone, to produce the sleep in highly susceptible subjects, after having been impressed. Hearing, and sight, and smell would require to be annulled before I could readily believe in community of taste and feeling, which all will easily comprehend, who have closely observed the powers of suggestion, in giving such vividness to ideas as shall delude the patients, even in the wide-waking state, and make them mistake mere ideas for realities."—"Magic, Witchcraft," etc., p. 82.

[1] *Medical Times*, April 12, 1844.

somnambulism." The following important case was cited in illustration of this condition.

"Mr Jones, eighteen years of age, applied to me at the beginning of last winter. About two years previously he had been attacked by a powerful dog, and was thrown violently on the ground, which hurt his back considerably, and alarmed him greatly, although there was no wound inflicted on his person by the teeth of the dog. Shortly after this he was attacked with fits of an epileptic character, which went on increasing in frequency and severity, in defiance of the medical treatment prescribed. Previously to applying to me he had frequently been out of one fit into another, for whole nights together, and had sometimes six or seven attacks in a day besides. His sleep was thus quite destroyed—sometimes eight or nine nights passed entirely without sleep. When Mr Jones first applied to me, I hypnotised him in the ordinary mode, but without apparent improvement. I now prescribed medicines for him *in addition* to the hypnotism, but without benefit. He was, therefore, sent into the country, but returned without improvement. On the 6th of January last I resolved to try hypnotism alone, and *in that mode calculated to depress the circulation*, and immediately the beneficial effects were most manifest. The first night he had no fit, next day only one, and from that day they declined both in force and frequency. On the fifth night he had a slight attack, and none since, being a respite of more than ten weeks. In order to test the extent of double consciousness in this case, before he entered the room, I wrote a letter to him, and having explained my intention to a scientific friend (without the patient's knowledge), in the presence of my friend after the patient was asleep, I read over the note, requesting him to repeat it after me, which, after the second reading he did, *with verbal accuracy*, excepting the date, which he placed *after*, instead of *before*, the month. I considered this of no material consequence, and therefore deposited the letter in a drawer, and in a little time after aroused the patient, who remembered nothing of what had been done, said, or heard, during his sleep. The following day, after being asleep and passed into the same stage as the previous day, the letter was taken out of the drawer, and held at a distance by one gentleman, whilst myself and another who was present the day previous, listened attentively to hear what he would say, when asked to repeat the letter he had from me the day before. He repeated it *with verbal accuracy*. Shortly after he was aroused, when I asked if he remembered anything of what had been said or done in his sleep, he remembered nothing. I then asked if he received a letter from me the day before, to which he replied in the negative, and moreover, added that he had never in his life received a letter from me. I told him that during his sleep he had admitted so, and even repeated its contents. He looked surprised at this, and could only be made to believe I was not hoaxing him by the other two gentlemen assuring him it was true. I then gave him the letter to read, and told him he had repeated the very words he now read, as a letter which he had received from me the day before. . . . After he had returned the letter to me, I requested him to repeat it now that he had read it whilst awake ; but he could not repeat a single line, only a few of the leading words, and the general purport of the letter. He never saw the letter from that moment; but on every occasion when asleep since, he has repeated it with verbal accuracy, but remembers nothing of it when awake. . . . Mr Jones called on me on the 11th March, after an absence of five weeks. He was quite well, and when hypnotised, repeated the above letter with verbal accuracy, but remembered nothing of it after being aroused."

In the *Medical Times* for Jan. 13th, 1844, Braid returned to the subject of Phreno-Magnetism and explained that "one of the peculiar features of the excitability of the nervous system induced by hypnotism" was "that the mind is liable to manifest itself as entirely absorbed in whatever individual passion or emotion it may be directed to; and, moreover, that an idea being excited in the mind, associated with contact with *any* part of the body, whether head, trunk or extremities, by *continuing such contact*, the mind might be rivetted for an almost indefinite length of time *to the same train of ideas*, which would work themselves up into more and more vivid manifestation, according to the length of time afforded for that purpose. It thus appeared to me that, by availing ourselves of these peculiarities, we might very readily determine the *relative forces of the different emotions and propensities*. For example, by exciting the various passions and emotions in succession, through auricular suggestion, and by fixing each new idea by mechanical contact with the *same point* of the patient, as the *suggestion and fixation of the ideas* would be the *same in all*, provided an equal length of time was allowed for each to develop the force of its manifestation, any *difference* in the *relative force* could only be attributable, on phrenological principles, to a corresponding difference in original development as

regards size, or to greater or lesser activity of the respective organs, arising from the degree of exercise of
corresponding portions of the brain. Then, again, by
comparing the *relative forces* of the *manifestations realised by such experiments*, with the *known character* of
the individual ; provided they both coincided, by again
comparing them with what a practical phrenologist
should determine, *simply from the cranial developments*,
as to what *ought* to be the character of such individual ;
if the latter was found to compare with the former, then
would hypnotism be a proof in favour of phrenology ;
but, if *they differed*, it would prove just as much *the
contrary.*"

The attempt to supplement the resources of a feeble
vocabulary by the incessant use of italics usually leads
to confusion, and the particulars of this laboured scheme
are a little difficult to grasp on a first reading. The
phreno-magnetism of Braid was the weakest part of his
theory, and though he approached it in the careful and
scientific spirit which governed his other researches, it
has failed to obtain recognition. In the plan under
notice it is easy to see that there were large sources
of error against which it was impossible to guard, the
" known character " of the individual being perhaps the
most obvious. However, the experiments took place,
in the presence of a number of scientific friends whom it
will be needless to cite by name. The heads of five
patients were examined by a phrenologist of Manchester, his reports were sent in, and remained un-
opened till the close of the proceedings. Each of the
subjects was then separately hypnotised, and all the
leading emotions denoted by phrenological organs were
successively excited.

" I considered," says Braid, " that putting a ring on
the same finger or thumb of a patient was the most
convenient mode of fixing the ideas suggested.   This,
therefore, was the mode of procedure.   On each occasion I spoke aloud to this effect :—Now, gentlemen, you
will see that the moment I place this ring on his [or her]
finger or thumb, as the case might be, you will observe
that he [or she] will think of *devotion*, at the same time
putting the ring on the finger or thumb indicated.   Immediately the emotion was manifested ; and, having
afforded it a certain time to develop itself, its *relative
force* was determined on,[1] and accurately noted on a
blank form.   The ring, being removed, the patient, who
had been kneeling, now arose ; " and another idea was
then excited and fixed in the same manner, the relative
force determined, the process continuing until all the most
important emotions and propensities had been tested.
The same plan was adopted with all the subjects.

The results obtained by the process were compared
with the phrenological record, but " in no instance did
the manifestations accord with the value or force set
upon the several organs, in more than four out of thirteen leading characteristics."   The result was therefore
unfavourable to phrenology.   When, however, the known
character of the patients was compared with the charts
of the delineator there was " a remarkable coincidence."
Whether the reading was erroneous in the one case or
the method of computing the relative force mistaken in
the other must be determined according to the predilec-
tion of the few persons interested in the question.   Nor

---

[1] One would like to know on what principles.   The oracular description
suggests some arbitrary method, or one at least which offered little
guarantee in its results.   See, however, Appendix V., No. 11.

will it be necessary to delay long over the observations offered by Braid in connection with his experiment. He affirms that mere size is insufficient to determine the force of function, much of the perfection of which may depend on the perfection of structure, and this again may be influenced by practice and habit.[1] Phrenology determines character according to the relative sizes of particular parts of the brain, and not according to the degrees of activity of such part, independently of size. "The manifestations realised by mesmerism and hypnotism, on the contrary, display the *energy acquired from habit or practice*" rather than the original proclivity. For all practical purposes this objection would appear to destroy phrenology, the follower of which is recommended to have recourse to hypnotism as a valuable adjuvant which will furnish him with a key for determining "how far habit, practice, and other concurring circumstances have been at work in aiding or counteracting original predisposition."

With these experiments Braid's connection with the debated subject of Phreno-Hypnotism seems to have terminated, though at a later period he promised to recur to it again and to provide a full explanation of his theory concerning its phenomena. The subject remains at the present day at the point where he left it. The verdict of Dr Foveau de Courmelles will perhaps most recommend itself to the impartial. "Such phenomena," he observes, "may as easily be considered the result of a real action as that of unconscious suggestion.[2]

[1] This is practically the conclusion previously reached by the author in "Neurypnology."

[2] The most recent defendant of Phreno-Magnetism is Mr James Coates, in his work on "Human Magnetism." He, however, does little more than collect the favourable opinions of previous writers and record his own personal conviction.

§ 6. ARCHÆOLOGY OF HYPNOTISM.—BRAID'S EMBROILMENT WITH ELLIOTSON.

At the close of 1844, Braid, who had been pursuing his researches into the question of the antiquity of hypnotism, published some articles as a result in the *Medical Times*, under the title of " Magic, Mesmerism, Hypnotism, etc., Historically and Physiologically considered," which further developed the suggestion already advanced that the marvels attributed to the fakeers were to be explained in the light of his discovery. He supposes that the secret of the induction of an intense form of abstraction through a strain on the eyes and suppression of the respiration, passed into India from Persia, whence also it was acquired by the Greeks. He regards the success of Orientals in auto-hypnotism ages before animal magnetism and mesmerism were heard of as a confirmation of his view that " there is no *magnetic fluid* or *special influence* requisite to produce the sleep, and many of the attendant phenomena." Into his disquisition upon mesmerism as the basis of magic art in ancient Egypt, its perpetuation in modern Egypt, its use among the Romans, and, in a word, its universal diffusion, there is no need to follow him. He lived to see the subject far more fully, comprehensively, and satisfactorily treated by J. C. Colquhoun, and Braid's articles, it must be confessed, are somewhat meagre as regards their materials and ill-balanced in their arrangement. This was probably realised by their compiler, and he re-edited their best portions, which were incorporated with his " Observations on Trance " and some later pamphlets.

Prior both to Braid and Colquhoun, the same subject had been treated, as one would have thought, most exhaustively by the German writer Ennemoser, whose " History of Magic" became so well known in England through the translation of William Howitt. The explanation of all past transcendentalism by the phenomena of animal magnetism is fundamentally perhaps not much more convincing than its treatment by the method of Salverte.

It has been said that the *Zoist* ignored Braid for years, and indeed from the foundation of that periodical until the death of the discoverer of hypnotism it swerved from its rule of silence only on one or two occasions, and then acted unamiably. The first occasion was the treatment of a young lady mesmerically by Elliotson subsequent to hypnotic treatment by Braid. The particulars of the squabble are not worth recording. Elliotson's defined hypnotism after the happy manner which characterised him as " the coarse method practised by Mr Braid," and Braid replied with spirit, carried the war into the enemy's own camp, and returned victorious so far as the *tu quoque* argument may be said to vindicate or establish anything. He did better than this by exhibiting the real state of the case which he had partially cured—a complication of acute spasm, spinal curvature, and afterwards contraction of the foot —when the patient went out of his charge, and was the subject of various experiments, sometimes by another physician, sometimes by her mother, as a result of which she suffered a relapse with some further complications. Dr Elliotson seems to have completed what was begun by Braid, and with any grace or good taste might have done so without giving just cause for offence.[1]

[1] In the year 1854, the publication of Townshend's " Mesmerism proved True and the Quarterly Reviewer Reviewed," gave another oppor-

§ 7. BRAID AND REICHENBACH. — COUNTER - EX-PERIMENTS CONCERNING THE NEW IMPONDERABLE.

Almost simultaneously with Braid, Baron von Reichenbach, an Austrian nobleman, living at the Castle of Reisenberg, near Vienna, was pursuing independently researches of another kind but also dealing with the alleged force of magnetism. The discovery which he was convinced that he had made had a different history from that of the English surgeon, and the contrast offered by the two is not an uninstructive episode in the history of scientific progress. There was everything in favour of the continental investigator. He was a man of birth and fortune, with leisure for scientific inquiry, and the advantage of a thoroughly scientific education, while Braid had only the ordinary opportunities of a college training in Scotland, and was probably dependent on a profession which absorbed most of his time. Reichenbach already occupied a recognised scientific position, which is said even to have been eminent,[1] while Braid was nothing more than a fairly

tunity to the *Zoist*, because Braid and his method were mentioned in that work. Townshend's "Facts in Mesmerism" was termed philosophical by his admirers, but this pamphlet appears to have been fantastic in argument and whimsical in style. The criticisms of the mesmeric journal, which dwell upon the fact that Dr Carpenter was the "dear friend" of "the renowned Mr Braid of Manchester" are conceived in as ill a spirit as they are written in unmannerly language. In the same year, however, the *Zoist* concluded its course, owing to the death of Elliotson. The possibility of better auspices made its demise regrettable.

[1] "He is known as a distinguished improver of the iron manufacture of his native country. He is a thoroughly practical chemist, and by his well-known researches on tar acquired a very high position. In geology, physics, and mineralogy, he has been equally active. In particular, he is the highest living authority on aerolites, of which remarkable bodies he possesses a magnificent collection."—Howitt's "Gleanings in the Corn Fields of Spiritualism." It should be added also that Reichenbach was the discoverer of paraffin and of creosote.

successful practitioner in the north of England; in a word, he was quite unknown. And yet the discovery of the Englishman has placed the subject of his researches in the rank of official science while the lapse of more than sixty years has scarcely materially advanced the claims of the Austrian's invention. The nature of these claims is, however, so well known as scarcely to need description. As the result of a long series of experiments with a number of sensitive subjects, Reichenbach announced the existence of "flame-like emanations from crystals, from the poles of a magnet, from the bodies of the sick, and from newly-made graves." The subjects further gave evidence of "various sensations experienced from contact with magnets and metals." The author of the experiments was led from these phenomena to infer the existence of a force, "the material nature of which is to us, as yet, as completely occult and hidden as it is enigmatic, as much so as that of light, electricity, and of other dynamics." While it differs from these, it is still associated with every substance in existence; in other words, it is of universal diffusion. Braid was the first serious English critic of the researches of Reichenbach, which gave occasion to his pamphlet, "The Power of the Mind over the Body: An Experimental Inquiry into the Nature and Cause of the Phenomena attributed by Baron Reichenbach and others to a New Imponderable." It was published in 1846, shortly after the annotated translation of Reichenbach issued by Dr Gregory.[1]

After what manner the experiments of the Austrian investigation were received in continental circles we

[1] Its substance had previously appeared in the *Medical Times* for June 13, and onward, of the same year. See Appendix V.

may learn at first hand from a letter to Dr Langsdorff, published many years ago in the American *Banner of Light*. He says: "I was received by the public with joyful shouts; the 'Letters,' in three editions, were devoured and translated into all the languages of Europe; but all in vain. The obstinate materialists like Liebig, Dubois, Vogt, and Schleiden, angrily assailed, without refuting, me, or even venturing an attempt to refute me. I retorted sharply, and since then they have kept silence. But I have now all these gentry for deadly enemies, and as their influence is omnipotent, every effort which I might make, to gain a reception for my doctrines, must be suppressed. . . . I have done what a man could; may courageous successors, in times more favourable, follow in my footsteps, and complete what my contemporaries have rendered it impossible for me to accomplish." Certainly in some instances the method of continental criticism sufficiently condemned itself. For example, Dubois Reymond, a Berlin doctor, described the experiments of Reichenbach as "an absurd romance, to enter into the details of which would be fruitless, and to him impossible," a judgment without examination, which might justly excite the anger of a peculiarly laborious investigation. It must be remembered, however, that twenty-three physicians of Vienna held no less than twenty-two sittings to test the experiments of their countryman and arrived at an unfavourable conclusion.

The criticisms of Braid proceeded, however, upon entirely different lines. He also repeated the experiments, but in expressing divergent views does every honour to the high reputation of their original describer, and opposes his authority with a diffidence which does honour to himself. "The great aim of Baron Reichenbach's researches in this department of science has been to establish the existence of a new imponderable, and to determine its qualities and powers in reference to matter and other forces, vital and inanimate." But on first reading the account of his "Researches on Magnetism," Braid says : "My experience with hypnotic patients enabled me to perceive a source of fallacy, of which the Baron must either have been ignorant, or which he had entirely overlooked." Whether ignorance or oversight, his critic did not accuse him of "any want of ordinary caution, or even the strictest desire to guard against sources of fallacy of every sort. . . . It is obvious that he had taken great precautions, and such, indeed, as would have been quite adequate to effect such purpose under ordinary circumstances. . . . A better devised series of experiments . . . for determining the question on strictly inductive principles, I have never met with." The source of fallacy in question was "the important influence of the mental part of the process," which was affirmed to be "in active operation with patients during such experiments." Premising that the test of the alleged new force is the human nerve, and that even so it can only be demonstrated to "a comparatively small number of highly sensitive and nervous subjects," he develops his objection as follows : "It is an undoubted fact that with many individuals, and especially of the highly nervous, and imaginative, and abstractive classes, a strong direction of inward consciousness to any part of the body, especially if attended with the expectation or belief of something being about to happen, is quite sufficient to *change the physical action of the part, and to produce such impressions from this cause alone, as Baron Reichenbach attributes to his new force.* Thus every variety of feeling may be excited from an *internal* or *mental* cause—such as heat or cold, pricking, creeping, tingling, spasmodic twitching of muscles, catalepsy, a feeling of attraction or repulsion, sights of every form or hue, odours, tastes, and sounds, in endless variety, and so on, according as accident or intention may have suggested. Moreover, the oftener such impressions have been excited, the more readily may they be reproduced, under similar circumstances, through the laws of association and habit. Such being the fact, it must consequently be obvious to every intelligent and unprejudiced person, that no implicit reliance can be placed on the human nerve, as a test of this new power in producing effects from *external* impressions or influences, when precisely the same phenomena may arise from an *internal* or *mental* influence, when no external agency whatever is in operation."

So as to guard against this source of fallacy, Braid in his counter-experiments threw his patients into the "nervous sleep," and operated on those whom he knew had no use of their eyes when in that state. He then observed the results of a magnet "capable of lifting fourteen pounds being drawn over the hands and other parts of the body, without contact," as described by Baron Reichenbach. "The results were, that in no instance was there the slightest effect manifested, unless when the magnet was brought so near as to enable the patient to feel the abstraction of heat (producing a sensation of cold), when a feeling of discomfort was manifested, with a disposition to move the hand, or head, or face, as the case might be, *from* the offending cause. This indication was precisely the same when the armature was attached, as when the magnet was open ; and in both cases, if I suffered the magnet to *touch* the patient, instantly the part was hurriedly withdrawn, as I have always seen manifested during the primary stage of hypnotism, when the patients were touched with any *cold* object. Now, inasmuch as patients in this condition, generally, if not always, manifest their perceptions of external impressions by the most natural movements, unless the natural law has been subverted by some preconceived notion or suggested idea to the contrary, and as I have operated with similar results upon a considerable number of patients, we have thus satisfactory proof that there was no real attractive power of a magnetic or other nature, tending to draw the patient, or any of his members, so as to cause an adhesion between his body and the magnet, as between the latter and the iron, as Baron Reichenbach had alleged. I conclude, therefore, that the phenomena of apparent attraction manifested in his cases were due entirely to a *mental* influence."

It is entirely outside the present purpose to offer observations on an argument which it is solely intended to present, but it may be noted in passing that Braid's experiments are not palpably conclusive, even so far as they go, and that their range seems singularly restricted. If his patients were not constitutionally disposed towards the peculiar quality of sensitiveness which proved rare in the subjects of Reichenbach, it does not absolutely follow that the indirect sensitiveness of the hypnotic sleep would provide the required conditions, and it further does appear that slight sensations were experienced by some of them, though they are referred by Braid to a more ordinary, though still an unaccountable, cause.

The alleged emanations from magnets, crystals, and other objects are approached, though also not altogether in a convincing manner. Braid lays stress upon the discrepances in the evidence of the sensitives, as to the colour, size, and situation of the flames, which were differently given by different subjects, and even by the same subjects at various times. " If there be a physical reality," he argued, " in these alleged flames and colours, there ought to be no discrepances of this sort ; and the fact of such discordant statements having been made will tend to prepare the reader for the problem which I am now about to submit." It is obvious, however, that a radiating force may not always manifest with the same degree of intensity, and that its variations may depend upon conditions which escape conjecture at the beginning of an inquiry into a subject which is wholly unknown.

" In my experience," says Braid, proceeding to his solution of the problem, " no light or flames have been perceived by patients either from the poles of a magnet, crystals, or other substance, unless the patients have been previously penetrated with some idea of the sort, or have been plied with such questions as were calculated to excite notions, when various answers were given accordingly, and when in the sleep, there appeared an equal aptitude to see something *when neither magnet nor fingers were in the direction indicated, as when they were*—a clear proof that the impressions were entirely imaginary, or mental in their origin." The solution is enforced by an account of some admirable experiments with subjects in the vigilant state which abundantly demonstrate that hallucinations of any given kind can be produced by suggestions on the part of the operator. At the present day the records of similar experiments constitute a considerable body of literature, and though they scarcely require to be fortified by revivals from the past, some of Braid's instances are so acute in their arrangement that they deserve to be remembered, and some accordingly will be found in an appendix at the end.[1] Before, however, they can be regarded as conclusive evidence against the objective nature of the phenomena observed by Reichenbach, it must be at least shown that such phenomena have not been recorded under circumstances when the possibility of suggestion could not be reasonably assumed. Many persons who are acquainted with the inception of Reichenbach's own experiments will probably rest persuaded that the facts which urged him to inquiry are not covered by Braid's theory, while the later evidence for the influence of magnets and metals on hypnotic patients, though it has been explained after the same manner in the light of counter-experiments analogous to those of the Manchester surgeon, the latter remain inconclusive, now as then, on account of their negative character.

In concluding his thesis, Braid drew particular attention to one point. " My experiments and researches," he said, " are opposed to the *theoretical* notions of Baron Reichenbach and the mesmerists," but, " in all the more important points, they directly confirm the reality of the facts, as to the power which we possess of artificially producing certain phenomena by certain processes ; as also of intensifying effects which arise in a minor degree, spontaneously, or by the patient's own unaided

[1] See Appendix III.

efforts. They allege that the exciting cause is the impulsion into the body of the patient from without of a portion of this new force; whilst I attribute it to a subjective or personal influence, namely, that of the mind and body of the patient acting and reacting on each other in a particular manner, from an intense concentration of inward consciousness on one idea, or train of ideas, which may, to a certain extent, be controlled and directed by others. The latter power, however, merely arises from the mental and physical impressions producing still greater concentration of the patient's attention in a particular direction; that is to say, by concentrating their attention to the point over which they see anything drawn, or upon which a mechanical, calorific, frigorific, or electric impression is made, whereby a greater supply of nervous influence, blood, and vital action, is drawn to the part *from the physical and mental resources of the patient himself,* and *not* from the person or substance exciting those physical impressions. They enable the patient more effectually to concentrate his own vital powers, and thus to energise function; on the same principle as a patient afflicted with anæsthesia, or loss of feeling, is able to hold an object in his hand whilst he looks at it, but will allow it to drop when his eyes are averted.

"It is worthy of particular remark that my researches prove the power of concentration of attention, as not only capable of changing physical action, so as to make some patients, in the wide waking state, imagine that they see and feel from an *external* influence what is due entirely to an *internal* or *mental* cause; but I have extended the researches, so as to prove, that the same law obtains in respect to all the other organs of special sense, and different functions of the body. My theoretical views, therefore, instead of diminishing, rather enhance the value of this power as a means of cure. They strikingly prove how much may be achieved by proper attention to, and direction of, this power of the human mind over the physical frame, and *vice versa*, in ameliorating the ills which flesh is heir to. . . . In the experiments of Baron Reichenbach, and the mesmerists generally, all which I have endeavoured to prove as requisite for the production of the phenomena referred to, is necessarily brought into play during their processes, *in super-addition to their alleged mesmeric fluid, or new force ;* of the latter, therefore, under such circumstances, I maintain that we have, as yet, no direct and satisfactory proof; and it is unphilosophical to attribute to a new and extraneous force what can be readily accounted for from the indipendent physical and psychical powers of the patient, which must necessarily be in active operation, along with their processes, and alleged new imponderable."

It does not appear that the intelligent criticism of Braid, thus attempting to create a *via media* between verified facts and extreme theories, evoked any response from Baron Reichenbach,[1] or even produced much impression in uncommitted circles. A compromise was probably distasteful alike to believers and

[1] Professor Gregory replied briefly, and transmitted a copy of his answer to Reichenbach. (See *Phrenological Journal,* Oct. 1846.) He gives examples of cases in which the possibility of suggestion was carefully avoided. The subject was resumed in the succeeding issue of the same periodical, that of Jan. 1847, when Professor Gregory, who had not heard from Reichenbach, mentions a meeting with Braid, at which he witnessed many of the latter's experiments. He considered them exceedingly interesting, but not warranting the interpretation placed on the researches of the Austrian scientist.

sceptics, and the scientific feeling of the day was certainly against the "Researches." When the work, as translated by Gregory, was ultimately, that is to say, in the year 1850, reviewed in the volumes of the *Lancet*, the policy of the notice was not, however, one of actual hostility but reservation of judgment, with particular regard to the perilous position occupied by author and translator as believers in the "unscientific marvels" of Animal Magnetism. But it discoursed of experiments which had been known for five years in England as if they were novelties of the hour, and as if criticism were yet to be provided with some material for judgment. The material provided by Braid was totally ignored, which reflects badly either upon the equipment of the reviewer or upon his bias.

By the year 1860 the feeling in medical and scientific circles as regards the odylic force of Reichenbach would appear to have changed considerably. William Howitt, writing in that year, says : " In the late discussions on Spiritualism every one must have remarked how kindly and completely the medical and scientific opponents have taken to Reichenbach and what they call his Od force. It is the last new principle which they have accepted, and by that they are now swearing lustily. . . . Would any one believe that this discovery has been as violently assailed . . . as Spiritualism and Spiritualists are now ? . . . The history of the ridicule and the abuse of the odyle force [1] is a perfect *facsimile* of the history

---

[1] Dr Ashburner was the first writer in England to show the inaccuracy of the term *odylic*, which was never used by Reichenbach, but was devised by Dr Gregory in his summarised version of the Austrian nobleman's now historic " Researches." The term used in the original was Od. There was an acrid controversy in the pages of the *Spiritual Magazine* as to whether the variation of the translator was accepted or not by the author,

of Modern Spiritualism."[1] William Howitt, whose forcible vocabulary made him the Cobbett of Modern Spiritualism, though a man of conspicuous and varied ability, was not a person of any scientific attainments, and it is to be questioned whether he represented accurately the scientific opinion which he mentioned. I think Mr Podmore better reflects the actual position when he says that " on the evidence of Reichenbach's prolonged and laborious researches the existence of this supposed magnetic sense obtained a certain degree of credence."[2] However this may be, Braid's minute pamphlet on " The Power of the Mind over the Body " is one of the most interesting of his publications, it contains some of his most sagacious experiments, and is in all respects entitled to the extended notice which has been given it in this place.[3]

§ 8. The Phenomena of Human Hibernation.

Braid's attention seems next to have been devoted to a subject which had previously attracted him, namely,

who himself propounded an etymology of the authentic word, deriving it from a Germanic root traced to the Sanscrit *vâ*, which is highly fanciful, but as *vâ* is connected with the Teutonic *vahen* (*vehen*), the Latin *vado*, and the Norse Vada, whence, says Reichenbach, Wodan, Odan, and Odin, " signifying the all-pervading power," we get at least the intention of its discoverer. But the most curious thing is that Éliphas Lévi, the French writer on magic, who knew nothing, except by hearsay, of the investigations of the Austrian scientist, traces the term to Kabbalism wherein Od is said to be the name given to the force of the Astral Light when regarded in its positive aspect, the passive form being called Ob, and Aour the equilibrium between them.

[1] *Spiritual Magazine*, Dec. 1860.

[2] " Apparitions and Thought-Transference," p. 374.

[3] The reputation of the writer was much increased by its publication, after which he tells us that he received numerous letters " from some of the most eminent members of the profession and of general science (*sic*), expressive of their entire concurrence with my views of the nature and cause and extent of mesmeric phenomena ; and also of my mode of explaining the extraordinary phenomena adduced by Baron Reichenbach, as proof of a new imponderable."—*Medical Times*, Feb. 27, 1847.

the alleged performances of Indian fakeers, who suffered themselves to be buried alive for days and weeks
together, sometimes in ordinary graves, sometimes in
sealed boxes.   His experience in hypnotism persuaded
him that these feats might be something more than the
result of collusion, and might indicate ability in the
operators to "throw themselves into a state of catalepsy
or trance, more or less profound, in which condition, like
the hybernating animals, all the vital functions are
reduced to the minimum of what is compatible with
continued existence and restoration to their former
activity," the feat being accomplished, as he explains, by "suppressing the respiration and fixing the
mind, just as was manifested by the well-attested case
of Colonel Townsend in this country, and by many
patients whom I have myself witnessed, who have
acquired the like power in a minor degree."

Some observations on the subject were published, as
already seen, by Braid in the *Medical Times*, and there,
and so also in other organs of the profession, and by
circulars, and so forth, he invited information with
a view to extend the evidence.   In this manner he
elicited from Sir Claude Wade the now well-known
narrative of the fakeer who was entombed at Lahore in
1837, another of minor interest and less evidential value,
contributed by Sir C. E. Trevelyan, and one of a similar
character, to which also much currency has been given,
though the officer of the East India Company on whose
evidence it rests, remains unknown.   The three cases
were incorporated by Braid into a small volume, called
" Observations on Trance, or Human Hybernation,"
which appeared in 1850.   The significance of the
analogy afforded by the habits of hybernating animals

is developed in this work with the writer's usual
sagacity, and the lessons of the narrative are illustrated
by original experiments. Some of these will be found
among the additional cases in the appendix. He had,
however, an experience of hypnotism in his own person
which may be reproduced in this place.

  " In the middle of September, 1844, I suffered from a
most severe attack of rheumatism, implicating the left
side of the neck and chest, and the left arm. At first
the pain was moderately severe, and I took some medicine to remove it ; but, instead of this, it became more
and more violent, and had tormented me for three days,
and was so excruciating, that it entirely deprived me of
sleep for *three nights successively*, and on the last of the
three nights I could not remain in any one posture for
five minutes, from the severity of the pain. On the
forenoon of the next day, whilst visiting my patients,
every jolt of the carriage I could only compare to several
sharp instruments being thrust through my shoulder,
neck, and chest. A full inspiration was attended with
stabbing pain, such as is experienced in pleurisy. When
I returned home for dinner, I could neither turn my head,
lift my arm, nor draw a breath, without suffering extreme pain. In this condition I resolved to try the
effects of hypnotism. I requested two friends, who
were present, and who both understood the system, to
watch the effects, and arouse me when I had passed
sufficiently into the condition ; and, with their assurance
that they would give strict attention to their charge, I
sat down and hypnotised myself, extending the extremities. At the expiration of *nine minutes* they
aroused me, and, to my agreeable surprise, *I was quite
free from pain, being able to move in any way with perfect*

ease. I say *agreeably* surprised, on this account: I had seen like results with many patients; but it is one thing to *hear of pain*, and another to *feel it.* My suffering was so exquisite that I could not imagine anyone else ever suffered so intensely as myself on that occasion; and, therefore, I merely expected a *mitigation*, so that I was truly agreeably surprised to find myself *quite free from pain.* I continued quite easy all the afternoon, slept comfortably all night, and the following morning felt a little *stiffness*, but *no pain.* A week thereafter I had a slight return, which I removed *by hypnotising myself once more;* and I have remained quite free from rheumatism ever since, now nearly six years."

The fact that the cataleptic condition, more especially when induced by disease, is not always accompanied by loss of consciousness is made the subject of certain inferences which are not of less importance at the present day than when Braid first drew attention to them. He lays down: 1st, "That great discretion ought to be used in applying tests for the purpose of determining the extent of torpor of the senses which may exist in any particular case, because it is quite possible that the patient may suffer intensely from such inflictions, without being able, at the time, to give any sensible sign of doing so, from the existence of a complete paralysis of all voluntary power"; 2nd, "That all tests which would produce laceration or permanent injury of tissue should be avoided, even in that class of patients in whom loss of feeling is complete, or who remember nothing of it afterwards, because, although not felt by the patient at the time of infliction, nor remembered afterwards, whether the state of loss of feeling may be the consequence of

spontaneous trance, or anæsthesia induced by hypnotism, mesmerism, chloroform, haschisch, or any other
narcotic, such inflictions are certain to manifest their
natural baneful influence on the patient, after he is
restored to the normal state of sensation " ; 3rd, " That
in all cases of trance, and indeed in *every case of death*,
no patient ought to be interred until there have been
indubitable signs of decomposition of the body,
because, without this appearance, there is no certain
sign of death."

From these inferences the author turns, finally, to the
best modes of treating catalepsy or trance so as to obviate the possibility of incurring the dangers to which
he has alluded. Premising that such cases occur
usually in females, and are closely allied to hysteria,
which again has an intimate relation with irregularities
in the female functions, he shows by some cases within
his own experience that " the catamenial function and
various morbid phenomena connected therewith, are so
remarkably under the control of the hypnotic mode of
treatment, that there can be no doubt whatever of its
superiority over all other methods in dealing with such
affections." [1]

§ 9. Hypnotism and Electro-Biology.

About this time the experiments of " Electro-
Biology " were imported from America. Under a
name which has been condemned not unjustly for its
pretentiousness and absurdity, they were concerned with
a matter which Braid recognised as of importance,
namely, suggestion in the waking state. The original

[1] See Appendix III., Note 4.

deviser was J. Stanley Grimes, who is usually said to have been unacquainted with the discovery of Braid. The subject fell, in the country of its origin, into the hands of extravagant theorists who used the vague term in the interests of tawdry speculation and explained the whole universe in bombastic lectures.[1] A typical representative of this worthless school was one John Bovee Dod, whose work on the " Philosophy of Mesmerism and Electrical Psychology" had an enormous sale on both sides of the Atlantic among uncritical persons.[2]

Braid, who was keenly, though not unduly, jealous of attempts to annex his discovery, was more than inclined to regard Electro-Biology as a piracy, and its alleged independence of his own researches would rest, naturally enough, under considerable suspicion if there were no evidence to the contrary. It would be curious indeed if the lapse of five or six years should have been insufficient for the report of hypnotism to have crossed the Atlantic, but as the priority of Braid is in any case unassailable, the matter is in itself, and especially at this day, of no importance. Durand de Gros, who appears to have been not only a careful observer but a most judicious thinker, remarks on this point " that without contesting the originality of electro-

[1] Much the same view was taken in France, so far back as 1850 by Durand de Gros, who observes : " Electro-Biology has been propagated in the United States by a multitude of professors, most of whom were not at the apex of a scientific mission."

[2] Grimes was himself an enthusiast and affords some curious reading. He appears to have begun as a phrenologist and produced a new system of that ill-starred science. A chart in accordance was published in 1843. In 1851 this system received an astonishing development when Grimes discovered that the organisation and functions of the brain were connected with the successive geological periods. In 1891 he was still discussing and disclosing the " Problems of Creation."

biology as a discovery, but because this discovery was posterior by many years to that of hypnotism, it seems reasonable to regard the two processes as progressive stages of one and the same art, of which the paternity rests with Braid." As a matter of fact, however, most writers who have concerned themselves with the question have not had recourse to the works of the chief exponent of Electro-Biology. Grimes in his " Philosophy of Mesmerism," a work which also passes by its title as Etherology, the Phreno-Philosophy, etc., not only makes use of the term Hypnotism, but mentions Braid, to whose principles and facts he denies the merit of novelty, not on account of any rival claim of his own, but on the ground that they are old Knowledge.

One of the earliest propagators of Electro-Biology in England seems to have been Dr Darling, whose notoriety is said to have overshadowed for a moment the unobtrusive fame of Braid. He was followed or accompanied by an American named Stone, who gave experiments all over the country, and among other places at Manchester, exciting a very large degree of interest among intelligent and scientific persons, including many of the foremost converts to Hypnotism. In June, 1851, Braid was moved to contribute an article on the phenomena in question to the *Monthly Journal of Medical Science*, in which he observes that " vigilant or waking phenomena" had been long known among mesmerists, and were in all respects similar to those of Electro-Biology, and that the zinc and copper discs of the American process were merely a variety of his own method for inducing the hypnotic state. They were "merely visible and tangible objects for aiding the

D

patients in fixing their attention, and inducing that state of mental abstraction which is the real origin and essence of all that follows." The difference between the processes is, however, clearly recognised as centred in the fact that Electro-Biology made mental and physical impressions on the subjects before passing them into the state of sleep. Braid, notwithstanding, maintained that he had performed and published similar experiments, namely, those in connection with the researches of Reichenbach, so far back as the year 1846. But the point which he was most solicitous to enforce was not the priority of discovery,[1] but the acting cause in the phenomena. Were they mental, or were they electric? Having established that the new experimentalists described aloud, within the hearing of their patients, the idea or act which they designed should be manifested, or otherwise resorted to some kind of visible signal or tangible impression, he regarded this as a clear proof that the influence was psychical or mental, not physical or electric. "Were it *electric*, there ought to be no more need for their auricular suggestions and manœuvres to produce the results on their patients than there would be for . . . the attendant to speak aloud to the electric telegraph the message which he intends to convey." The alleged instances of the silent and unseen communication of behests to subjects, through pure sympathy and power of will, were too few in Braid's opinion to sustain the electric hypothesis. "At best they are but exceptional cases ; and when I have had opportunities of testing such patients, I have soon been able to de-

[1] During his visit to Manchester, Stone, at his own request, was introduced to the discoverer of hypnotism, who showed him the experiments recorded in the " Power of the Mind over the Body," to the great surprise of the electro-biologist.

monstrate the sources of fallacy which had misled the operators." But Braid's most curious argument is this : "It is a well-established physiological fact, that the moment the trunk of a nerve is divided, all sense and voluntary motion are abolished in parts supplied by such nerve beyond the point of section. Now are we to believe it possible that such sympathy could exist between different individuals as to enable one, by the exercise of his volition, or other silent and unseen manœuvres, to force his own nervo-vital influence beyond his own organism, so as to control the acts of the other at a distance, even miles apart, when he could not propel it a fraction of a line beyond the section of the nerve in his own limb (the divided ends of the nerve being in close apposition), so as to produce a voluntary movement of his own member ? "[1]

Having sought to establish the mental cause of the phenomena, Braid proceeded to show that they were not produced by a mere play upon the will of the patient and the influence of sympathy and imitation. These might be sufficient to account for the phenomena in which muscular motion was concerned, but not for the larger and more surprising class, the suspension, intensification, or perversion of impressions addressed to any or all of the organs of special sense. These results could be accounted for only by the power of an over-excited and vivid imagination, and fixed idea,

---

[1] Without espousing the hypothesis of the transmission of nervo-vital influence between operator and subject, it is permissible to say that Braid's crux for electro-biology is of the high fantastic order. Operator and subject are both living and active stations or centres of the nervous force, and its "silent and unseen" transmission from living station to living station may be impossible, but no argument against it can be founded on the false analogy of a communication which has been severed.

changing physical action—a view which was already
familiar to the readers of Braid's pamphlets.

§ 10. CONTROVERSY WITH J. C. COLQUHOUN.

Braid was not only a believer in the power of the
mind over the body, but he regarded his experiments
as proofs of the unity of the mind; he was also, as
appears, by his books, a Christian of orthodox views.
The correspondence of Harriet Martineau and H. G.
Atkinson, a book which at one time excited more
interest than it deserved, but has been long since forgotten, moved him almost to indignation on account of
some worthless opinions on religion which appeared in
it; while on one occasion he went out of his way to
maintain that the existence of a personal devil could not
be called in question "without rejecting the obvious
interpretation of various important parts of scripture."
It would seem unlikely that such a person should be
accused of materialism.   This, however, appears to have
been a charge preferred against Braid by the author of
Isis Revelata, J. C. Colquhoun.   It was rather by im-
plication than otherwise, and anyone less anxious to
preserve the character of a strictly orthodox thinker
would have been scarcely at the pains of replying to it.
It was, in any case, replied to, and, as one would think,
somewhat too warmly and without duly considering
whether the words of the criticism really bore the con-
struction placed on them.   Mr Colquhoun was among
the first to deny that they did, and to make an honourable recantation, which Braid gives gladly enough in
an appendix to his published justification.   This was
entitled " Magic, Witchcraft, Animal Magnetism, Hyp-

notism, and Electro-Biology," and for some inscrutable reason seems to have passed through three editions,[1] successively enlarged, the last of which, under date of 1852, being designed to "furnish a periscope or vidimus" of the author's views "on all the more important points of the hypnotic and mesmeric speculations." It is precisely for this reason, namely, that it contains little which is new, that it can be dismissed with a short description. There is no interest in its controversial and personal part, and the rest is largely a resumption. There is indeed one matter on which it may be worth while to dwell for a moment. Braid was strongly indisposed to accept any evidence for clairvoyance. He was a plain man, without any intellectual subtlety, and with no gift for metaphysics, and he did what most plain-thinking men would have done with him in those days, and perhaps not a few even now ; he jumped to the conclusion that to " see through opaque bodies " was a gift possible only to " omniscience," and he therefore regarded the pretension as " a mockery of the human understanding." This view, which finds expression in the " Observations on Trance," was not unnaturally the object of Colquhoun's hostile criticism, and Braid, whose good sense was unteachable, reiterated and defended it. Moreover, he had tried the clairvoyants, and had found them only gifted with the discernment of things which did not exist. For him the phenomena of the supposed faculty were simply " a dream spoken and acted out, directed and modified by suggestions of those present," and partaking of the peculiar character of dream, " which is to accept every idea arising in the mind, or suggested to it through impressions on the senses, as present

[1] It should be observed that the first two are entirely unknown in England.

realities, with a tendency, as in insanity, to reason correctly, occasionally, from the erroneous premises which
had been assumed as true.   Like dreaming and fortune-
telling, moreover, the answers are given in very vague
phraseology, so that they may admit of any variety of
interpretation which may best suit the fancy of the
parties interested in the issue of the inquiry." Again :
" The *whole* of the really striking achievements of mesmeric clairvoyants—which are not simply occasional
coincidences and shrewd guesses—are merely results of
concentrated attention, quickened memory, exaltation of
the natural organs of special sense, with self-confidence,
and accurate deductions as to what might be calculated
upon regarding the future, from contemplation of the
circumstances in the existing case, compared with what
was known from past experience. . . . In fine, they are
conclusions drawn from given premises, when contem-
plated with a quickness of perception, conception, and
memory, and with a force of reason, unknown to us
during the ordinary diffused, and distracted, or ever-
varying state of our attention during the waking
condition."

The case against clairvoyance had been put somewhat
differently by the author, so far back as the year 1843,
in a paper contributed to the *Medical Times*. Premising that " one of the most interesting phenomena connected with hypnotism is that extraordinary activity of
the imagination, whereby ideas excited on the mind,
whether from recalled past impressions, or by oral
suggestion or otherwise, are instantly invested with all
the attributes of reality," he accounts in this manner for
the elaborate descriptions of places and persons, so frequently given by subjects who are under a genuine

impression that they actually behold what they are describing. With these observations on clairvoyance, the reader may compare what was said by Braid at a still earlier date on the subject of the transposition of the senses, which in the case of the faculty of sight connects obviously with the same subject, and the two classes of alleged phenomena have sometimes been confused.[1]

In taking leave of this controversial pamphlet, it may be noted : (1) That Braid claims for hypnotism a capability of achieving all and more than all the good attainable by mesmerising processes, while more speedily inducing the condition. (2) That he himself in practice used both modes indifferently. Lastly, he says : "One decided advantage in favour of hypnotism is this—that I have never experienced the slightest difficulty in arousing any of my patients, even from the most profound stage of the sleep which they may pass into ; nor do they experience any annoyance from other persons than the operator approaching to, or touching them, during the sleep—both of which inconveniences frequently manifest themselves, in a marked degree, in the ordinary mesmeric state ; for some mesmerised subjects will be thrown into the most violent agitation, with intense catalepsy or convulsions, merely by another person touching them—which is called cross-mesmerism —or they may remain for many hours, and sometimes even for *days*, in the sleep, without being able to be aroused. I consider this merely arises from a predominant idea or conceit in the mind of the patient, and [the fact of] the operator not knowing the key to the puzzle, presents the great difficulty of unlocking him from the condition."

[1] See " Neurypnology," p. 117, *note.*

§ 11. LAST WRITINGS OF JAMES BRAID.

In the year 1853, Braid contributed a paper to the *Monthly Journal of Medical Science*, which was immediately reprinted, with an appendix, in a separate pamphlet, under the title of " Hypnotic Therapeutics," and is a summarised statement of his entire case for hypnotism as a healing agent. It traverses a second time the ground covered by many previous pamphlets and papers, and occupies the same position towards this portion of his subject that was fulfilled towards hypnotism as a whole, at least in the intention of its writer, by the preceding publication. Being free, however, from the personal and, it must be confessed, somewhat puerile controversy of " Magic, Witchcraft, Animal Magnetism," etc., it is a better piece of work, but, as in that case, does not present much material which calls for description. The quoted cases are for the most part not new and there is nothing fresh in the aspect. Beyond the notes which have been derived from this pamphlet for the first and second appendix, it is only necessary to observe that " Hypnotic Therapeutics " seems to prove, what is clear also from his other writings, that, despite the entire separation of Hypnotism from Animal Magnetism which Braid announced in the introduction to " Neurypnology " he really never definitely made up his mind as to this question. Indeed the separation which he originally set up was only one of method, and his terms are much too pronounced for the real nature of his distinction. In " Hypnotic Therapeutics " he repeats his disbelief as to the reality of any magnetic fluid, nervous, or vital force, passing from operator to patient, but at the same time admits that mesmeric

passes do exercise an influence, and again states that
he resorted to them on account of their mechanical
impression on the subject. Possibly his position was
clear in his own mind, but he was incapable of expressing himself lucidly over any difficult subject. The
erroneous appreciations of his position in respect to
Animal Magnetism, which have appeared from time to
time, are largely the direct consequence of his uncertain
and confused mode of writing.

Braid's next pamphlet was published in 1855, under
the title of " Electro-Biological Phenomena, and the
Physiology of Fascination." It divides with the " Letter
to a Clergyman " the distinction of being the most obscure and unprocurable of his writings. The only copy
known to the present editor is in the Library of the
Manchester Medical Society at Owens College, but it
did not escape the patience and research of Preyer, who
has included it in his translations of Braid, together
with its appendix, or second part, entitled " The Critics
Criticised." Its connection with the marvellous had
recommended the term Fascination to American biolo-
gists, and its philosophy had been duly set forth. It
will be unnecessary to say that in Braid's hands the
romantic element evaporates, and that he treats under
this name the peculiar condition sometimes induced by
great unexpected dangers both in men and animals, as
also of its artificial production in the former by occasioning the abnormal predominance of one idea. Fascination
of this kind he distinguishes from catalepsy, which is
the loss of will but not of motion. The absorption of
the entire attention by a single conception produces
now the simple hypnotic state, unaccompanied by
rigidity, now the rigidity of catalepsy, and again fascina-

tion, which have all in common a temporary disturbance of the activity of the nervous centres originating in the excitement of the prevailing idea. On the one hand, a complete suspension of motive power may be occasioned, as when a nervous person crossing a crowded thorough-fare is fascinated by the apprehension of danger, so that he cannot leave the spot ; on the other hand, the arrested attention may send an impulse through nerves and muscles, causing a corresponding movement without any control of the will, and then such a person is com-pelled to advance, or seems drawn, into the very heart of the danger. This explains for Braid not only the attraction of the bird to the snake, when astonishment first takes the attention and then fear dynamises the muscles, but also the phenomena of table-turning, which, he observes, have deceived many, "who thought the tables drew them, but it was they who pushed the tables." In France at the same period Velpeau ac-counted for the raps which succeeded the turning by a slight cracking of the muscles in the calf of the leg—those raps, says Éliphas Lévi, the French transcenden-talist, with quiet scorn, which sometimes break tables and almost demolish walls.

The critics who are subjected to Braid's criticism in the addendum to this pamphlet are the Rev. George Sandby, once well known by his work on " Mesmerism and its Opponents," and Townshend, author of " Facts in Mesmerism," also of considerable repute.

On one occasion Braid magnetised a blind woman suffering from pains in the chest caused by a wound ; he relieved her of this pain by passes made over the spot without contact. On another occasion the same patient was affected beneficially, when suffering from rheumatism in the knee-joint, by passes made along the feet at a distance of one yard. Mr Sandby challenged Braid to show that these results could be produced in a blind subject merely by fixed attention, and Braid replied by quoting from his work on "Neurypnology" the remarks which appear at p. 113 of this edition on experiments with persons deprived of the sense of sight. Mr Townshend's objection turned upon the necessity of the operator's presence, from which he inferred the fact of the operator's influence, and Braid in his answer shows that on the occasion of his first public lecture he caused three of his patients to pass into the hypnotic state while he remained in another room, which he quitted only on ascertaining that they had fallen asleep without any other influence than their own concentrated attention. This was witnessed by six hundred spectators.

The pamphlet on Fascination was followed in the same year by a small work on the treatment of certain forms of paralysis by the agency of hypnotism, no appliances and no medicines being used. It is a synopsis of cases only, some of which had appeared already in "Neurypnology." Of the others a typical instance will be found in the second appendix, and may be held to suffice for all. This closed the literary labours of James Braid.[1] His publications were all of the character which is called occasional; they give evidence that he wrote with difficulty and probably with dislike for composition; with a meagre vocabulary and little skill, his work could not have been a pleasure, but there is, of course, no jurisdiction for its criticism from the standpoint of literary excellence. His death, it may be

[1] See Appendix IV.

added, attracted little attention in scientific circles. The obituary notice which appeared in the *Manchester Guardian* on the following day, though appreciative, was singularly short. The concluding lines may be quoted, because they embody all that is likely to be known as to the social life of the Manchester surgeon in the city with which he was identified for so many years of his life. It says that Braid had much more than a local reputation, and that this was due "not alone to his theory of hypnotism" but in a very large degree "to his special skill in dealing with some dangerous and difficult forms of disease. Amongst his friends he was ever warm-hearted and cheerful; and his kindness in devoting his skill for the benefit of humble sufferers, in cases where money recompense could not be anticipated, has endeared him to large numbers by the strongest feelings of gratitude."

His surgical skill was the subject of special mention in the *Lancet* of March 31, 1846. "Long before his discovery of hypnotism, he had performed some extraordinary cures by operations on contracted muscles, in cases of club-foot and similar contortions, which brought him patients from all parts of the kingdom." The obituary notice containing this passage also admits that Braid applied hypnotism in certain diseases with great effect.

§ 12. PROGRESS OF HYPNOTISM AFTER THE
DEATH OF BRAID.

Though Braid died without much apprehension as to the verdict of posterity on his discovery, and though he had received a measure of recognition from the sources which he may be supposed to have valued most, he had only gathered the first fruits of the harvest due to him, and though he had no means of knowing it, he was actually at the time of his death on the threshold of a fuller reward. Perhaps, indeed, he may have anticipated it to a certain degree, for the coming event had begun to cast its shadow in his direction.

In the *Comptes Rendus* of the French Académie des Sciences, t. 49, No. 23, being the weekly issue for Dec. 5, 1859, there is a note by M. P. Broca on the subject of a new anæsthetic method, being an account of Braid's discoveries, and of experiments by the writer in conjunction with B. F. Azam.[1] This was apparently brought to Braid's notice, and he sent most of his pamphlets to the Academy, the receipt of which was acknowledged. They were presented by M. Velpeau, with whom he had been probably in correspondence, as he had been certainly with Azam.[2] It is recorded that they were accompanied by a MS. note, "in which the author seems to have summed his observations on these singular nervous states." M. Velpeau was invited by the Academy to make himself acquainted with the publications in question, and to submit a verbal report if needed. Of this report, if made, there is no record in the "Proceedings" subsequently published. There is fortunately, however, a valuable document of

---

[1] A month later Azam himself contributed a "Note on the Nervous Sleep, or Hypnotism" to the *Archives Générales de Médecine*, which is partially reproduced in the writer's subsequent work. It deals with Braid and his researches in a very friendly spirit, even as regards phreno-hypnotism, the experiments of which were not successfully reproduced by the critic.

[2] The work of Dr Azam, published many years after, *Hypnotisme, Double Conscience, et altérations de la Personnalité*, does not mention the fact, though it makes many references to Braid, to whose writings the author was introduced by Bazin.

the period from which the effect of Velpeau's interest may be ascertained. This is the "Theoretical
and Practical Course of Braidism,"[1] which was published by Durand de Gros under the name of Philips.
It appears from the preface of that work that a most
extraordinary sensation was created and the peace-
able, even cordial, acceptation of the alleged facts by
the Académie des Sciences is explained by the supposition that both the institution and its informant
Velpeau had been entrapped, and had given welcome
to an old acquaintance in the disguise of a new comer.
There is reason, however, to believe that Velpeau at
least was quite well acquainted with the connection
existing between Animal Magnetism and Hypnotism.[2]
Du Gros goes on to state that the merchandise so long
prohibited was soon freely commerced in under its
novel trade-mark, and that under the pretext of Hypnotism official scientists could devote themselves to the
study of the marvellous, without risking either their
appointments or their professional consideration. The
most accredited medical journals gave the place of
honour to communications for which at another period

[1] *Cours Théorique et Pratique de Braidisme, ou Hypnotisme Nerveux
considéré dans ses Rapports avec la Psychologie, la Physiologie, et la
Pathologie, etc.* Par le Docteur J. P. Philips. Paris, 1860.
[2] That he had many opportunities for knowledge is shown by the
Bulletin de l'Académie Impériale de Médicine. In vol. xxv., No. 7,
under date of Dec. 27, 1859, it is recorded that Dr Sandras submitted
to the judgment of the Company a note on hypnotism and its dangers,
when the Commissioners nominated to report were Velpeau, Bousquet,
and Jolly. On Jan. 10, 1860, some information on hypnotism was addressed to the Academy by M. Philips, and the same Commissioners were
named. Again, on Jan. 17, M. Delfraysse offered an exposition of a new
method invented by him for the induction of the hypnotic state. Commission as before. According to Binet and Féré, Velpeau appeared to
have no doubt that Animal Magnetism, which had been condemned by
the Academy, had reappeared under a new name.

they would have gladly committed the writers to an asylum.

Unfortunately the mode of the moment had either the usual fate of a nine days' wonder or the Academy found that it had been entrapped.[1] In either case, the subject of hypnotism in France had to pass through a certain ordeal of hostility, though it was not of long endurance, and not of an implacable character. The existence of the nervous sleep was admitted on all sides, and any radical opposition was therefore practically impossible.

At the close of an introduction which has already reached its utmost limits a monograph on the progress of hypnotism in France is of course out of the question, and there are otherwise a variety of handbooks, historical and practical, which have popularised information on the subject. It may be said in a general manner that until the year 1878, there was little attempted beyond the verification of Braid's researches and variations of his mode of inducing the hypnotic state. Though admirable in many respects, and possessing something of a literary accent which still makes them pleasant reading, the lectures of Durand de Gros did little to advance the science. He sought to perpetuate its association with the English discoverer by adopting the term Braidism in preference to hypnotism, " because just as animal magnetism implies a debated theory, so hypnotism makes sleep the essential and

---

[1] Binet and Féré represent it as a natural indifference to a tedious and by no means infallible method which was put aside in favour of the convenience and certainty of chloroform. " The year 1860," they say, " witnessed," for the moment, " the dawn and decline of the prevailing fashion of employing hypnotism to produce surgical anæsthesia." Their account, however, is exceedingly sketchy and the dates are somewhat mixed.

constant character of the phenomena," and he appealed
to the precedents established by galvanism, voltaism,
etc.   The word, however, was awkward and the one
devised by Braid in the end obtained the preference.
For the rest, the *Cours du Braidisme* is academic rather
than practical, and concerned largely with what has
been called the "mechanism of hypnosis."   Between
the concentration of thought induced by fixity of gaze
and the manifestation of insensibility, catalepsy, extasis,
and the profound revolution of the economy which takes
place in the subject, there was, said Du Gros, a dividing
gulf, and this he attempted to bridge by his doctrine of
ideo-plasticism and the hypotaxic state.   It was, however, too early for theorising, and Durand de Gros
seems to have attracted little attention.   He was followed by Liébault, who had at first no better success,
and by a number of writers who verified without extending the experiments of Braid.   The metalo-thera-
peutic theory, first discovered by Burq and developed
by Charcot, inaugurated a new era of hypnotism, and
led to that long series of researches connected with the
Salpêtrière Hospital of Paris, which have done so much
to obscure, if not indeed to tarnish, the rival lustre of
the Nancy School of Liébault.

The name of James Braid is a household word
among the hypnotists of both schools, and there is no
book, as there is indeed scarcely an article published in
French scientific periodicals, which has not referred to
him.   The father of hypnotism has indeed obtained
universal recognition and acknowledgment.   With the
one exception of Heidenhain,[1] the case is much the

---

[1] The late Professor Romanes in reviewing Heidenhain's *Der Sogen-
nante thierische Magnetismus*, suggests that Heidenhain was unacquainted

same in Germany. There is no need to insist on this point, and it remains only to indicate that he is at the present day in the position rather of a pioneer than a paramount authority, as must be almost invariably the case with the discoverer of a new science. At the same time, as in England so in France, there is an insufficient first-hand acquaintance with the writings of James Braid; his best works have not been translated; his views, and even his experiments, are known largely by tradition and at second hand. There are therefore one or two points upon which he seems to have suffered a little unjustly at the hands of modern criticism. A full acquaintance with his works would appear to make it indubitable that he distinguishes quite clearly the existence of various stages in the hypnotic state, but he has been charged with confusing them. This criticism did not perhaps originate with MM. Binet and Féré, but it has been popularised in their well known work, and in books by other writers, such as Fouveau de Courmelles, who have followed their lead. There is also an erroneous impression that Braid regarded fixity of gaze as essential to the production of the hypnotic condition, but this opinion, originally held by him, was abandoned after his successful operation on blind subjects. Finally, his experiments in phreno-magnetism have left him under a certain suspicion that he was a believer rather than an observer, but here also his later conclusions did something to modify what was unfounded in his earlier pronouncements. On the other hand, it has been rightly observed that Braid, the discoverer, so to speak, of suggestion, by no means fully realised the

with Braid's works, but it seems incredible that he should not have been acquainted at least with Braid's position as discoverer of Hypnotism. See *Nineteenth Century*, September, 1880.

E

possibilities of suggestion of the unconscious kind. His position in relation to animal magnetism has also been correctly appreciated by French writers, who have rectified the once prevailing error that he was in an active sense of the term an opponent thereof. He opposed indeed its infatuations, though also some of those "higher facts" which may not be less actualities because they still await their full and authoritative confirmation. These facts he was constitutionally incapable of appreciating. But in so far as Animal Magnetism, separated in his own mind from unaccredited theories, and from discredited marvels, was based, according to his judgment, upon natural truth, he was its champion and witness. Indeed, the greatest glory of James Braid is not that he discovered something new and unheard of, or revived something lost and forgotten, but that he established the reality of artificial somnambulism, placing the central fact of mesmerism beyond all further dispute, so that it became a "state subject to observation," which anyone can produce at pleasure.

ORIGINAL DEDICATION

TO

CHARLES ANDERSON, M.D., F.R.C.S.Ed., &c.

My Dear Sir,

Inclination as well as duty induces me to dedi-
cate this work to you : that I may publicly express the
lively sense of gratitude I entertain for the many opportunities enjoyed during my apprenticeship with yourself
and your late father, of acquiring a practical as well as
theoretical knowledge of my profession ; of becoming
familiar with your comprehensive views of disease, and
their happy application in practice ; for the personal
kindness shewn me during my pupilage ; and for the
uninterrupted friendship which has ever since existed
betwixt us.

I am aware that you are not practically acquainted
with the subject of this treatise, but your intimate
knowledge of general science renders you eminently
qualified for prosecuting the subject with success ; and

permit me to assure you, that its value as a curative power, for an intractable class of diseases, renders it well worthy of your best attention.

    Believe me,

        My dear Sir,

            Most faithfully yours,

                JAMES BRAID.

3 ST PETER'S SQUARE,
MANCHESTER, *June* 2, 1843.

AUTHOR'S PREFACE

THE circumstances which led me to engage in the investigation of hypnotism are detailed in the introduction to this little work; in the first part of the treatise I have endeavoured to give the results at which I arrived, in most instances sketching the route by which I travelled, and stating the inferences drawn from the various incidents which occurred in the course of my progress. Having furnished the data from which I drew my conclusions, the reader is thus prepared to determine, whether, on any occasion, I have come to these conclusions without what he would consider sufficient evidence, and in such case can institute additional experiments to any extent he may judge requisite. One circumstance, however, I may remark. From a fear of being misled, I have requested the most sceptical individuals I knew, both professional and merely scientific men, to scrutinize all my experiments in the most critical manner; and have also induced some of my most intelligent and respected friends to submit to the operations, in the hope that I might thus more certainly guard against being deceived. The results I now submit to the public, and to the kind and candid consideration of my professional brethren, whom I should wish to investigate the subject coolly, and with an honest desire to arrive at truth. Having myself been sceptical, I can make every reasonable allowance for others. On this

point I fully subscribe to the propriety of the remark of
Treviranus, the celebrated botanist, when speaking of
mesmerism. He says (I quote from memory), " I have
seen much which I would not have believed on your
telling ; and in all reason, therefore, I can neither hope
nor wish that you should believe on mine."

It is quite natural for any man to prefer the evidence
of his own senses to that of all others, and I think no
one who has the opportunity of examining the phenomena for himself should neglect to do so. However,
there are some circumstances which ought to be particularly borne in mind, or very erroneous opinions may
be formed by the uninitiated, from what is actually
witnessed. First, there is a remarkable difference in
the degree of susceptibility of different individuals to
the hypnotic influence, some becoming rapidly and intensely affected, others slowly and feebly so. This is
only analogous to what we experience in regard to the
effects of medicines on different individuals, and especially as regards wine, spirits, and opium, and nitrous
oxide. Whilst this is a recognized fact, as regards the
latter, it appears to me somewhat surprising to find
many, and even professional men too, who seem to
expect as much uniformity ought to obtain, in regard
to the phenomena during hypnotism, as if we were
operating on inanimate matter. On the contrary, they
ought to be ready to admit that a variety might be
expected to arise, even in the same individual, according to the physical and mental condition of the patient
at the moment the operation is performed.

The next most important point for consideration is,
the fact of all the phenomena being consecutive. We
have thus the extremes of insensibility, and exalted

sensibility, of rigidity and mobility, at different stages, and these merging into each other by the most imperceptible gradations, or in the most abrupt manner, according to the mode of treating the patient. It is no unusual thing for different parties to be testing, or calling for tests, for the *opposite conditions*, at the *same instant of time*. These, of course, are incompatible, but, at a certain stage, the transitions from torpor of all the senses, and cataleptiform rigidity, to the most exalted sensibility, and flaccidity of muscle, may be effected almost with the celerity of thought, even by so slight a cause as a breath of air directed against the part. If left at rest it will speedily merge back again, and thus those unacquainted with such peculiarities, will be continually liable to think they discover discrepancies, which, however, only originate from their imperfect knowledge of the subject ; just as an unskilful manipulator will be ready to suppose, from his different results, that the observations of other chemists have been erroneous.

The third point meriting especial attention is, the condition of the *mind* at different stages. As results from opium, so also from hypnotism. At one stage it gives an extraordinary power of concentration of thought, or disposition to rapt contemplation, whereas, at another stage, the discursive, or imaginative faculties are excited into full play, and thus the most expanded, bright, and glowing scenes and images are presented to the fervid imagination. Such effects are quite analogous to those described as resulting from the use of opium, and detailed by the late Sir Humphrey Davy, as experienced in his own person, from the inhalation of the nitrous oxide. " I thus felt a sense of tangible extension, highly pleasurable in every limb, my visible impressions were dazzling and apparently magnified. I heard distinctly every sound in the room, and was perfectly aware of my situation. By degrees, as the pleasurable sensation increased, I lost all connection with external things ; trains of vivid visible images rapidly passed through my mind. I existed in a world of newly connected and newly modified ideas."[1] It must also be borne in mind, that these opposite *mental* conditions may glide into each other by the most imperceptible degrees, or by the most abrupt transitions, according to the modes of management, and thus consciousness or unconsciousness, sound sleep, dreaming, or somnambulism, will result, according as sensations or ideas predominate, or are equally vivid. See Hibbert's *Philosophy of Apparitions.* At a certain stage the same abruptness of transition may be realised in the *mental* phenomena, as were referred to in the last paragraph, in respect to the *physical*, and from equally slight causes. I presume it is from this cause that the phrenological manifestation may be so readily and characteristically exhibited at this period. At page 210 I have stated that, were it not that I should consider it an unnecessary waste of time to prosecute the inquiry farther, after the amount of evidence obtained by myself, and others, I had no doubt but I might soon obtain any number of additional cases I might desire. In proof of this, I may remark, that since that period, I was induced one day to try some fresh subjects, when I succeeded in eliciting the manifestations in the most satisfactory manner in the case of a man of forty years of age, and in three other male subjects upwards of twenty years of age.

[1] See Appendix I., Note 1.

Of the latter, under the excitation of constructiveness and ideality, one wrote, and the other drew patterns, and neither of them had seen such experiments, nor expected to be so tested, nor remembered what happened. The same day I also manipulated three females, one 45 years of age, a young lady of 19, and a girl, all of whom exhibited the manifestations quite distinctly. Another day, to satisfy a number of intellectual friends, I hypnotized three of their personal friends, two of whom were entire strangers to me, and were quite sceptical as to the possibility of my being able to affect them at all. They all exhibited the manifestations most distinctly, two of them in a remarkable degree, and to the extent of twenty manifestations at first trial. Under "conscientiousness," one restored a reticule she had stolen, and burst into a flood of tears at the thought of her delinquency. The friends were alarmed at the intensity of her emotion, but by changing the point of contact, I had her changed from the grave to the gay in a few seconds. A few days after I had other two cases, and I feel assured that in most of the twelve cases here referred to the parties knew nothing of phrenology, and that not one of them could with certainty point to two of their own organs. I may also add, they were all tested before competent and observant witnesses, who can testify there was no prompting by any one.

It appears quite evident, that whatever images or mental emotions or thoughts have been excited in the mind during nervous sleep, are generally liable to recur, or be renovated and manifested when the patient is again placed under similar circumstances. Notwithstanding the apparent conclusiveness of the cases recorded, that there exists a *natural* connection betwixt certain local-

ities touched and the peculiar manifestations which follow, in order to determine this question in the most
decided manner, it is my intention to institute a series of
experiments on fresh patients, in order to ascertain to
what extent it may be practicable, by arbitrary association, to excite the *opposite* tendencies from the *same*
points ; also, whether they can be exhibited in the same
striking and natural manner by *both* methods, or by
which points they can be elicited with the greatest
facility and fidelity to natural expression. There will
thus be both positive and negative proof to aid us in
determining, whether there is any natural and necessary
connection existing betwixt the points manipulated, and
the manifestations excited ; or whether it may depend
entirely upon associations which have originated from
some partial knowledge of phrenology, from arbitrary
arrangements, or accidental circumstances or causes
which have been entirely overlooked or forgotten ; and
which afterwards produce the results from " that ultimate
law of the mind, which ordains, that the repetition of a
definite sensation shall be followed by a renovation of
the past feelings with which it was before associated."
(*Hibbert*, page 316.)   I am induced to adopt this course,
from my anxiety to remove every possible source of error
as to the cause of the original manifestation, and from
the recollection of the remarkable circumstance of the
woman who, during natural somnambulism, could repeat
correctly large portions of the Hebrew Bible, and other
books, in languages she had never studied, and was perfectly ignorant of when awake, but which was at length
discovered to have been acquired from hearing a clergyman, with whom she resided when a girl, reading them
aloud to himself ; and also of some patients whilst labour-

ing under disease remembering languages long forgotten. I wish to ascertain whether any such accidental circumstance may have been the cause of the remarkable manifestations arising in the minds of patients when first manipulated. Whatever are the results of my farther inquiry shall be carefully noted and published, as my object is neither to prove nor disprove the truth of phrenology, but to establish the value of hypnotism, and determine how best to apply it, as a means of meliorating the mental, and moral, and physical condition of man.[1]

That during the nervous sleep, there is the power of exciting patients to manifest the passions and emotions, and certain mental functions, in a more striking manner than the same individuals are capable of in the *waking* condition, no one can doubt who has seen much of these experiments. And it can in no way alter the importance of hypnotism, as a curative power, and extraordinary means of controlling and directing mental functions, in a particular manner, by a simple association of impressions, whether we thus act on the brain as a *single* organ, or as a combination of separate organs ; or whether the primary associations have originated from a special organic connection, or from some accidental and unknown cause, or from preconcerted arrangement and arbitrary association.

In such operations as particularly require the use of the eyes, I have never seen patients in the hypnotic state perform what they attempted with the same celerity and accuracy as they were capable of doing when awake, and with the aid of sight. In short, I have never witnessed any phenomena which were not

[1] See Biographical Introduction, § 5.

reconcileable with the notion that they arose from the abnormal exaltation or depression of sensations and ideas, or to their being thrown into unusual and varied ratios by the processes resorted to.

Having heard it reported, that by establishing a connection betwixt two patients by a chain or string, that manipulating one would manifest the same phenomena in both, I tried the experiment, but with the precaution that the patients should be in separate rooms, so that the one could not hear, nor feel, from the motion of the air, what the other was doing. I formed the connection by a cord in some cases, and in others by a copper wire, and had parties stationed where they could observe the movements of both at same time. We could discover no such sympathetic influence as is asserted to have been realized by others.

The experiments recorded at page 209 of my having caused patients to hypnotize, manipulate, and rouse themselves (by simply desiring them to rub their own eyes), and which produced results precisely the same as when done by any one else, seem to me the most decisive proof possible that the whole results from the mind and body of the patients acting and re-acting on each other, and that it has no dependence on any special influence emanating from another. My first experiments on this point were instituted in the presence of some friends on the 1st May 1843, and following days. I believe they were the first experiments of the kind which had ever been tried, and they have succeeded in every case in which I have so operated.

With due attention to the points above referred to, and with that practice which is requisite to insure adroitness in experimenting in any department of art or science, and with an honest desire to view every fact having no bias to uphold some previous prejudice or opinion, I have no doubt that the facts and observations set forth in this treatise will soon be very generally confirmed.

A perusal of the cases recorded in the second part will, I trust, render the importance of the subject sufficiently apparent to stimulate inquiry; and I hope it may be gratifying to others to read, as it is to me to be able to record, that the prediction I had ventured to make at page 316, as to the probability of hypnotism proving a cure for tetanus and hydrophobia, has already been happily realized in respect to the former intractable and generally fatal disease. After this treatise had been in the press the following case occurred, and its importance must be my apology for giving a brief detail of it here:—

Master J. B., 13 years of age, was suddenly attacked with chilliness and pain all over his body, on the evening of 30th of last March. I was called to attend him the following day, when I considered he had got a febrile attack from cold, and prescribed accordingly. Next day, however, it had assumed a very different aspect. I now found I had got a severe case of opisthotonos to deal with. The head and pelvis were rigidly drawn back, the body forming an arch, and the greatest force could not succeed in straightening it, or bringing the head forward. Whilst the spasm never relaxed entirely, it frequently became much aggravated, when the head was so much drawn back as to seriously impede respiration. The legs were also sometimes flexed spasmodically. The effect of the

spasm in obstructing the respiration, and hurrying the
circulation, was very great, and seemed to place the
patient in great jeopardy. The pulse was never less
than 150, but during the paroxysm was considerably
increased. It was evident I had got a most formidable
case to contend with, and that no time ought to be lost.
I therefore determined to try the power of hypnotism,
well knowing how generally such cases end fatally
under ordinary treatment. He was quite sensible, and
the only difficulty in getting him to comply with my
instructions, arose from the recurrence of the severe
spasmodic attacks. In a very few minutes, however, I
succeeded in reducing the spasm so that his head could
be carried forward to the perpendicular, his breathing
was relieved, his pulse considerably diminished, and I
left him in a state of comparative comfort. In about
two and a half hours after I visited him again, accompanied by my friend Dr Cochrane. The spasms had
recurred, but by no means with the same violence.
Dr Cochrane had no difficulty in recognising the
disease, but did not believe any means could save such
a case. He had never seen a patient hypnotized till
that afternoon, and watched my experiment with much
interest and attention. He seemed much and agreeably surprised by the extraordinary influence which
an agency so apparently simple exerted over such a
case. The pupil was speedily dilated, as if unde the
influence of belladonna ; the muscular spasm relaxed,
and in a few minutes he was calmly asleep. Having
ordered three calomel powders to be given at intervals,
we left him comfortably asleep. Next day there was
still spasm of the muscles, but by no means so severe.
Whilst I determined to follow up the hypnotic treat-

ment, which had been so far successful, I considered it would be highly imprudent to trust *wholly* to that in the treatment of such a case. As I consider such cases are generally attended with inflammation of the medulla oblongata, and upper part of the spinal cord, I bled him, and ordered the calomel to be continued. The same plan was persevered in, hypnotizing him occasionally for some days, administering calomel till the gums were slightly affected, cold lotion to the head, and the antiphlogistic regimen till I considered all risk of inflammatory action past, when he was treated more generously, and I am gratified to say he is now quite well.

I feel confident that without the aid of hypnotism this patient would have died. I sincerely wish it may prove equally successful in other cases of the kind, and also in that hitherto fatal disease, hydrophobia. My anxiety to see it fairly tried in the latter disease induces me to offer my gratuitous services in any case of that disease occurring within a few hours' journey of Manchester.[1]

I consider it necessary to explain that my reason for having inserted some cases attested by the patients and others is, that most unwarrantable interferences have been resorted to by several medical men, in order to misrepresent some of them. In one instance, in order to obtain an *attested erroneous* document, the case was READ to *the patient and others present*, THE VERY REVERSE OF WHAT WAS WRITTEN. However extraordinary such conduct may appear, the fact of its occurrence was *publicly proved, and borne testimony to by the patient and other parties present on the occasion when the document was obtained.*

[1] See Appendix I., Note 2.

F

INTRODUCTION

IT was my intention to have published my "Practical Essay on the Curative Agency of Neuro-Hypnotism," exactly as delivered at the Conversazione given to the members of the British Association in Manchester, on the 29th June 1842. By so doing, and by appending footnotes, comprising the data on which my views were grounded, it would have conveyed a pretty clear knowledge of the subject, and of the manner in which it had been treated. It has since been suggested, however, that it might readily be incorporated with the short Elementary Treatise on Neuro-Hypnology, which I originally intended to publish, and which I am earnestly solicited to do, by letters from professional gentlemen from all quarters. I now, therefore, submit my views to the public in the following condensed form. I shall aim at brevity and perspicuity; and my great object will be to teach others all I know of the modes of inducing the phenomena, and their application in the cure of diseases, and to invite my professional brethren to labour in the same field of inquiry, feeling assured, that the cause of science and humanity must thereby be promoted.

It was with this conviction I offered my "Practical Essay on the Curative Agency of Neuro-Hypnotism," to the medical section of the British Association.

In November 1841, I was led to investigate the pre-

tensions of animal magnetism, or mesmerism, as a complete sceptic, from an anxiety to discover the source of
fallacy in certain phenomena I had heard were exhibited
at M. Lafontaine's conversazioni. The result was, that
I made some discoveries which appeared to elucidate
certain of the phenomena, and rendered them interesting, both in a speculative and practical point of view.
I considered it a most favourable opportunity for having
additional light thrown upon this subject, to offer a
paper to the medical section of the British Association,
which was about to meet in Manchester. Gentlemen
of scientific attainments might thus have had an opportunity of investigating it, and eliciting the truth, un-
biassed by local or personal prejudice. I hoped to learn
something from others, on certain points which were
extremely mysterious to me, as to the *cause* of some
remarkable phenomena. I accordingly intimated my
intention to the secretaries, by letter, on 18th May, and
on the morning of Wednesday, the 22nd June 1842,
sent the paper I proposed reading for the consideration
of the committee, intimating also, by letter, my intention
to produce before them as many of the patients as
possible, whose cases were referred to in proof of the
curative agency of Neuro-Hypnotism, so that they
might have an opportunity of ascertaining, for themselves, the real facts of the cases, uninfluenced by any
bias or partiality that I might exhibit as the discoverer
and adapter of this new mode of treatment. The committee of the medical section, however, were pleased to
decline entertaining the subject.

Many of the most eminent and influential members
of the Association, however, had already witnessed and
investigated my experiments in private, and expressed

themselves highly gratified and interested with them. In compliance with the repeated desire of these gentlemen, and many other eminent members of the Association to whom I could not possibly afford time to exhibit my experiments in private, and who were anxious to have an opportunity afforded them of seeing, hearing, and judging of the phenomena for themselves, I gave a gratuitous conversazione, when I read the " Rejected Essay," and exhibited the experiments in a public room, to which all the members of the Association had been respectfully invited. The interest with which the subject was viewed by the members of the Association generally, was sufficiently testified by the number and high respectability of those who attended on that occasion ; in reference to which the chairman requested the reporters to put on record, "that he had been in the habit for many years of attending public meetings, and he had never in his life seen a more unmixed, a more entirely respectable assembly in Manchester." It was also manifested by their passing a vote of thanks at the conclusion of the conversazione, for my having afforded an opportunity to the members of the British Association of witnessing my experiments, to which they had previously borne testimony as having been "highly successful."

On that occasion I stated, there were certain phenomena, which I could readily induce by particular manipulations, whilst I candidly confessed myself unable to explain the *modus operandi* by which they were induced. I referred particularly to the extraordinary rapidity with which dormant functions, and a state of cataleptiform rigidity, may be changed to the extreme opposite condition, by a simple waft of wind, either from the lips, a pair of bellows, or by any other mechanical means. I solicited information on these points, both privately and publicly, from all the eminently scientific gentlemen who honoured me with their company during the meetings of the British Association in this town; but no one ventured to express a decided opinion as to the causes of these remarkable phenomena. I now beg to assure every reader of this treatise, that I shall esteem it a great favour to be enlightened on points which I confess are, at present, still above my comprehension.

It will be observed, for reasons adduced, I have now entirely separated Hypnotism from Animal Magnetism. I consider it to be merely a simple, speedy, and certain mode of throwing the nervous system into a new condition, which may be rendered eminently available in the cure of certain disorders. I trust, therefore, it may be investigated quite independently of any bias, either for or against the subject, as connected with mesmerism; and only by the facts which can be adduced. I feel quite confident we have acquired in this process a valuable addition to our curative means; but I repudiate the idea of holding it up as a universal remedy; nor do I even pretend to understand, as yet, the *whole range of diseases* in which it may be useful. Time and experience alone can determine this question, as is the case with all other new remedies.[1]

When we consider that in this process we have acquired the power of raising sensibility to the most extraordinary degree, and also of depressing it far below the torpor of natural sleep;[2] and that from the

---

[1] See Appendix I., Note 3.
[2] *Vide* Experiments, pp. 135 to 139.

latter condition, any or all of the senses may be raised to the exalted state of sensibility referred to, almost *with the rapidity of thought*, by so simple an agency as a puff of air directed against the respective parts ; and that we can also raise and depress the force and frequency of the circulation, locally or generally, in a most extraordinary degree, it must be evident we have thus an important power to act with. Whether these extraordinary physical effects are produced through the imagination chiefly, or by other means, it appears to me quite certain, that the imagination has never been so much under our control, or capable of being made to act in the same beneficial and uniform manner, by any other mode of management hitherto known.

That we really have acquired in this process a valuable addition to our curative means, which enables us speedily to put an end to many diseases which resisted ordinary treatment, I think will be satisfactorily manifested by the cases which I have recorded. Many of these cases have been seen by other medical men, and are so remarkable, so self-evident to every candid and intelligent mind, that it is impossible, with any show of propriety, to deny them. Most unwarrantable and novel attempts have been made, not only to extinguish the farther prosecution of Hypnotism, but also to misrepresent all I had either said or done on the subject, and thus damage me, as well as Hypnotism, in public estimation. I am in possession of a mass of documentary evidence in proof of this, to an extent which could scarcely be credited. But I shall not trrouble my readers with details of all that has been done in order to prejudice my patients against me.

As regards general principles, it has even been

attempted, by garbled statements, to set forth such
gross misrepresentations as could only be credited by
parties totally ignorant of the subject. Thus it was
alleged, that my mode of hypnotizing was no novelty;
on the contrary, that it was an unacknowledged plagiarism, and that it was the opinion and practice of
Bertrand and the Abbé Faria. Now, so far as I have
been able to comprehend the meaning of Bertrand,
which Colquhoun observes, "it is rather difficult to
comprehend," he adheres "to the theory of imagination,
and imagination alone," (Colquhoun's *Introduction*, p. 94).
At p. 34, vol. iv. of the "Encyclopædia of Practical
Medicine," Dr Prichard says of Bertrand, that he "comes
at last to the conclusion, that all the results of these
operations are brought about through the influence of the
mind;" that is, through the influence of the imagination
of the patients acting on themselves. Bertrand also
supports this opinion by the manner in which the Abbé
Faria performed magnetization. His plan was this:
" He placed the patient in an arm-chair, and after telling
him to shut his eyes, and collect himself, suddenly pronounced, in a strong voice and imperative tone, the
word 'dormez,' which generally produced on the individual an impression sufficiently strong to give a slight
shock, and occasion warmth, transpiration, and *sometimes* somnambulism." Had his success by this method
been as general as mine, would he have used the word
"*sometimes*" on this occasion?[1] It is farther added,
"if the first attempt failed, he tried the experiment a
second, third, and even a fourth time, after which he
declared the individual incapable of entering into the
state of lucid sleep." Whilst it is doubted that his

[1] *Vide* pp. 105, 106.

success was equal to what he represented it, still Bertrand states, in reference to the Abbé Faria, that it was incontestable, "that he very often succeeded." Now, is this not sufficient proof, that his success was by no means so general as mine? And who does not see, on perusing my directions for hypnotizing,[1] that our methods are very different?

It is farther added, "The complete identity of the phenomena thus produced by a method which operated confessedly through the imagination, with those which display themselves under the ordinary treatment of the magnetizers, affords a strong reason for concluding that the results in other instances depend upon a similar principle." It is still farther added, that M. Bertrand denies the necessity of strong intense volitions of the operator being necessary to produce the result. He declared, "that in trials made by himself, precisely the same results followed, whether he WILLED to produce them or not, provided that the patient was *inwardly persuaded that the whole ritual was duly observed.*" Can any farther remarks be required to prove that Bertrand referred the result entirely to the effect of imagination? And can any one who has attended to what I have given as my opinion, say that this either *was*, or *is* my opinion? Certainly quite the contrary. The parties referred to, therefore, have only proved their belief of how easy it is, by garbled statements, to misrepresent the truth, when submitting such remarks to those ignorant of the subject, or who are blinded by prejudice.

The following remarks by Mr H. Brookes, a celebrated lecturer on animal magnetism, will illustrate this point rather better than the individuals referred

[1] For proof of this, see pp. 109 and 110.

to. On hearing that I had changed my original opinion about *identity*, he writes thus: "I am very glad you have at length found reason to change your original opinion as to the identity of your phenomena with those of mesmerism. From the very first I freely admitted the value and importance of your discovery, but I could not admit that identity, and I blamed you for insisting upon it so hastily, and using such hard words against the animal magnetists, because they could not agree with you. I thought, and still think, you did wrong in that, and that you certainly did yourself injustice, for in fact you are the original discoverer of *a new agency*, and not of a mere modification of an old one."

But when so much had been said of Bertrand, with the hope of making it appear that I had either been ignorant of, or copied his views without 'due acknowledgment, which is evidently erroneous, why not have quoted him also to prove I was wrong in attributing curative effects as resulting from these operations? Let us hear what M. Bertrand says on this point. He "declares, that it is difficult to imagine with what facility the practisers of the art succeed in relieving the most severe affections of the nervous system. Attacks of epilepsy, in particular, are rendered considerably less frequent and severe by their method skilfully employed; which displays in so remarkable a manner the influence of moral impressions on the physical state of the constitution." After such declarations in favour of the *curative power* of mesmerism, had M. Bertrand's method of inducing the condition been as generally and speedily successful as mine, will any one believe that it would not have been brought more generally into practice ere now? Mr Mayo, one of our best authorities, in a letter to me on this subject, states distinctly that the great reason for its not being more generally introduced into practice, was the tediousness of the processes for inducing the condition, and the uncertainty, after all the time and trouble devoted to the manipulation, of producing any result whatever. He concludes his observations on this subject, by the remark, "*It took up too much time.*" And Dr Prichard, author of the article referred to in the Encyclopædia of Practical Medicine, adds, "On the whole, when we consider the degree of suffering occasioned by disorders of the class over which magnetism exerts an influence, through the medium of the imagination, and the little efficacy which ordinary remedies possess, of alleviating or counteracting them, it is much to be wished that this art, notwithstanding the problematical nature of the theories connected with it, were better known to us in actual practice."

I am aware great prejudice has been raised against mesmerism, from the idea that it might be turned to immoral purposes. In respect to the Neuro-Hypnotic state, induced by the method explained in this treatise, I am quite certain that *it* deserves no such censure. I have proved by experiments, both in public and in private, that during the state of excitement, the judgment is sufficiently active to make the patients, if possible, even *more* fastidious as regards propriety of conduct, than in the waking condition; and from the state of rigidity and insensibility, they can be roused to a state of mobility, and exalted sensibility, either by being rudely handled, or even by a breath of air.[1]

[1] See Appendix I., Note 4.

Nor is it requisite this should be done by the person who put them into the Hypnotic state. It will follow equally from the manipulations of any one else, or a current of air impinging against the body, from any mechanical contrivance whatever. And, finally, the state cannot be induced, in any stage, unless with the knowledge and consent of the party operated on. This is more than can be said respecting a great number of our most valuable medicines, for there are many which we are in the daily habit of using, with the best advantage in the relief and cure of disease, which may be, and have been rendered most potent for the furtherance of the ends of the vicious and cruel; and which can be administered *without the knowledge of the intended victim.* It ought never to be lost sight of, that there is the *use* and *abuse* of every thing in nature. It is the *use*, and only the *judicious use* of Hypnotism, which I advocate.

It is well known that I have never made any secret of my modes of operating, as they have not only been exhibited and explained publicly, but also privately, to any professional gentleman, who wished for farther information on the subject. Encouraged by the confidence which flows from a consciousness of the honesty and integrity of my purpose, and a thorough conviction of the reality and value of this as a means of cure, I have persevered, in defiance of much, and, as I think, unwarrantable and capricious opposition.

In now unfolding to the medical profession generally —to whose notice, and kind consideration, this treatise is more particularly presented—my views on what I conceive to be a very important, powerful, and extraordinary agent in the healing art; I beg at once distinctly to be understood, as repudiating the idea of its being, or ever becoming, a universal remedy. On the contrary, I feel quite assured it will require all the acumen and experience of medical men, to decide in what cases it would be safe and proper to have recourse to such a means; and I have always deprecated, in the strongest terms, any attempts at its use amongst unprofessional persons, for the sake of curiosity, or even for a nobler and more benevolent object—the relief of the infirm ; because I am satisfied it ought to be left in the hands of professional men, and of them only.[1] I have myself met with some cases in which I considered it unsafe to apply it at all ; and with other cases in which it would have been most hazardous to have carried the operation so far as the patients urged me to do.[2]

In now submitting my opinions and practice to the profession in the following treatise, I consider myself as having discharged an imperative duty to them, and to the cause of humanity. In future, I intend to go on quietly and patiently, prosecuting the subject in the course of my practice, and shall leave others to adopt or reject it, as they shall find consistent with their own convictions.

As it is of the utmost importance, in discussing any subject, to have a correct knowledge of the meaning attached to peculiar terms made use of, I shall now give a few definitions, and explain my reasons for adopting the terms selected.

Neurypnology is derived from the Greek words νεῦρον,

---

[1] See Appendix I., Note 5.
[2] The circumstances which render my operations dangerous, the symptoms which indicate danger, and the mode of acting when they occur, to remove them, are pointed out, pp. 128 and 129.

nerve ; υπνος, sleep ; λογος, a discourse ; and means the
rationale, or *doctrine* of *nervous* sleep, which I define to
be, " a peculiar condition of the nervous system, into
which it can be thrown by artificial contrivance : " or
thus, " a peculiar condition of the nervous system, induced by a fixed and abstracted attention of the mental
and visual eye, on one object, not of an exciting
nature." [1]

By the term " Neuro-Hypnotism," then, is to be
understood " nervous sleep ; " and, for the sake of
brevity, suppressing the prefix " Neuro," by the terms—

HYPNOTIC,		The state or condition of *nervous* sleep.
HYPNOTIZE,		To induce *nervous* sleep.
HYPNOTIZED,		One who has been put into the state of *nervous* sleep.
HYPNOTISM,	Will be understood,	*Nervous* sleep.
DEHYPNOTIZE,		To restore from the state or condition of *nervous* sleep.
DEHYPNOTIZED		Restored from the state or condition of *nervous* sleep.
and		
HYPNOTIST,		One who practises Neuro-Hypnotism.

Whenever, therefore, any of these terms are used in
the following pages, I beg to be understood as alluding
to the discovery I have made of certain peculiar phenomena derived and elicited by my mode of operating ;
and of which, to prevent misconception, and interming-
ling with other theories and practices on the nervous
system, I have thought it best to give the foregoing
designation.

I regret, as many of my readers may do, the incon-

[1] See Appendix I., Note 6.

venient length of the name ; but, as most of our professional terms, and nearly all those of a *doctrinal* meaning,
have a Greek origin, I considered it most in accordance
with good taste, not to deviate from an established
usage. To obviate this in some degree, I have struck
out two letters from the original orthography, which
was Neuro-Hypnology.

NEURYPNOLOGY

PART I

CHAPTER I

HAVING, in the introduction, presented a cursory view of certain points, and given a few explanatory remarks, I shall now proceed to a more particular and detailed consideration of the subject. I shall explain the course I have pursued in prosecuting my investigation; the phenomena which I discovered to result from the manipulations had recourse to; the inferences I was consequently led to deduce from them; the method I now recommend for inducing the hypnotic condition, for applying it in the cure of various disorders, and the result of my experience, as to the efficacy of hypnotism as a curative agent.

By the impression which hypnotism induces on the nervous system, we acquire a power of rapidly curing many functional disorders, most intractable, or altogether incurable, by ordinary remedies, and also many of those distressing affections which, as in most cases they evince no pathological change of structure, have been presumed to depend on some peculiar condition of the nervous system, and have therefore, by universal consent, been denominated "*nervous complaints;*" and

G

as I felt satisfied it was not dependent on any special agency or emanation, passing from the body of the operator to that of the patient, as the animal magnetizers allege is the case by their process, I considered it desirable, for the sake of preventing misconception, to adopt new terms, as explained in the introduction.

I was led to discover the mode I now adopt with so much success for inducing this artificial condition of the nervous system, by a course of experiments instituted with the view to determine the cause of mesmeric phenomena. From all I had read and heard of mesmerism (such as, the phenomena being capable of being excited in so few, and these few individuals in a state of disease, or naturally of a delicate constitution, or peculiarly susceptible temperament, and from the phenomena, when induced, being said to be so exaggerated, or of such an extraordinary nature), I was fully inclined to join with those who considered the whole to be a system of collusion or delusion, or of excited imagination, sympathy, or imitation.

The first exhibition of the kind I ever had an opportunity of attending, was one of M. Lafontaine's conversazioni, on the 13th November 1841. That night I saw nothing to diminish, but rather to confirm, my previous prejudices. At the next conversazione, six nights afterwards, *one* fact, the inability of a patient to *open his eyelids*, arrested my attention. I considered that to be a *real phenomenon*, and was anxious to discover the physiological cause of it. Next night, I watched this case when again operated on, with intense interest, and before the termination of the experiment, felt assured I had discovered its cause, but considered it prudent not to announce my opinion publicly, until I had an opportunity of testing its accuracy, by experiments and observation in private.

In two days afterwards, I developed my views to my friend Captain Brown, as I had also previously done to four other friends; and in his presence, and that of my family, and another friend, the same evening, I instituted a series of experiments to prove the correctness of my theory, namely, that the continued fixed stare, by paralyzing nervous centres in the eyes and their appendages,[1] and destroying the equilibrium of the nervous system, thus produced the phenomenon referred to. The experiments were varied so as to convince all present, that they fully bore out the correctness of my theoretical views.

My first object was to prove, that the inability of the patient to open his eyes was caused by paralyzing the levator muscles of the eyelids, through their continued action during the protracted fixed stare, and thus rendering it *physically* impossible for him to open them.[2] With the view of proving this, I requested Mr Walker, a young gentleman present, to sit down, and maintain a fixed stare at the top of a wine bottle, placed so much above him as to produce a considerable strain on the eyes and eyelids, to enable him to maintain a steady view of the object. In three minutes his eyelids closed,

---

[1] By this expression I mean the state of exhaustion which follows too long continued, or too intense action, of any organ or function.

[2] Attempts have been made to prove, that I got this idea from a person who publicly maintained that the patient referred to *could* have opened his eyes *if he liked ;* to this the patient having replied, " I have tried all I could and cannot ; " the individual referred to, in support of his opinion, alleged, that the inability *was only imaginary ;* that he "could easily believe that a man may stand with his back to a wall, and may really believe that he has no power to move from the wall." It is therefore clear this individual attributed the phenomenon to a *mental,* whilst I attributed it to a *physical* cause.

a gush of tears ran down his cheeks, his head drooped, his face was slightly convulsed, he gave a groan, and instantly fell into profound sleep, the respiration becoming slow, deep and sibilant, the right hand and arm being agitated by slight convulsive movements. At the end of four minutes I considered it necessary, for his safety, to put an end to the experiment.

This experiment not only proved what I expected, but also, by calling my attention to the spasmodic state of the muscles of the face and arm, the peculiar state of the respiration, and the condition of the mind, as evinced on rousing the patient, tended to prove to my mind I had got the key to the solution of mesmerism. The agitation and alarm of this gentleman, on being roused, very much astonished Mrs Braid. She expressed herself greatly surprised at his being so much alarmed about nothing, as she had watched the whole time, and never saw me near him, or touching him in any way whatever. I proposed that she should be the next subject operated on, to which she readily consented, assuring all present that she would not be so easily alarmed as the gentleman referred to. I requested her to sit down, and gaze on the ornament of a china sugar basin, placed at the same angle to the eyes as the bottle in the former experiment. In two minutes the expression of the face was very much changed ; at the end of two minutes and a half the eyelids closed convulsively ; the mouth was distorted ; she gave a deep sigh, the bosom heaved, she fell back, and was evidently passing into an hysteric paroxysm, to prevent which I instantly roused her. On counting the pulse I found it had mounted up to 180 strokes a minute.

In order to prove my position still more clearly, I

called up one of my men-servants, who knew nothing of mesmerism, and gave him such directions as were calculated to impress his mind with the idea, that his fixed attention was merely for the purpose of watching a chemical experiment in the preparation of some medicine, and being familiar with such he could feel no alarm. In two minutes and a half his eyelids closed slowly with a vibrating motion, his chin fell on his breast, he gave a deep sigh, and instantly was in a profound sleep, breathing loudly. All the persons present burst into a fit of laughter, but still he was not interrupted by us. In about one minute after his profound sleep I roused him, and pretended to chide him for being so careless, said he ought to be ashamed of himself for not being able to attend to my instructions for three minutes without falling asleep, and ordered him down stairs. In a short time I recalled this young man, and desired him to sit down once more, but to be careful not to go to sleep again, as on the former occasion. He sat down with this intention, but at the expiration of two minutes and a half his eyelids closed, and exactly the same phenomena as in the former experiment ensued.

I again tried the experiment by causing Mr Walker to gaze on a different object from that used in the *first* experiments, but still, as I anticipated, the phenomena were the same. I also tried him *à la Fontaine*, with the thumbs and eyes, and likewise by gazing on my eyes without contact, and still the effects were the same, as I fully expected.

I now stated that I considered the experiments fully proved my theory; and expressed my entire conviction that the phenomena of mesmerism were to be accounted for on the principle of a derangement of the

state of the cerebro-spinal centres, and of the circula-
tory, and respiratory, and muscular systems, induced,
as I have explained, by a fixed stare, absolute repose
of body, fixed attention, and suppressed respiration,
concomitant with that fixity of attention.   That the
whole depended on the physical and psychical condition of the patient, arising from the causes referred to,
and not at all on the volition, or passes of the operator,
throwing out a magnetic fluid, or exciting into activity
some mystical universal fluid or medium.   I farther
added, that having thus produced the *primary* phenomena, I had no doubt but the others would follow as a
matter of course, time being allowed for their gradual
and successive development.[1]

For a considerable time I was of opinion that the
phenomena induced by my mode of operating and
that of the mesmerizers, were identical ; and, so far as
I have yet personally seen, I still consider the condi-

[1] It has been asserted, for the mere purpose of proving the contrary,
that I had claimed being the first to discover that *contact* was *not* necessary,
and that a magnetic fluid was not required to produce the phenomena of
mesmerism.   I never made any such claim, but illustrated these facts by
the most simple and conclusive experiments probably which were ever
adduced for that purpose.   In one of my lectures, I gave a history of
mesmerism, including Mesmer's attempt to mesmerize trees in Dr Franklin's garden, to prove to the Commission of 1784, that the patients would
become affected when they went under the mesmerized trees, from the
magnetic fluid passing from the trees to the patients.   This was proof
sufficient, that even *Mesmer* did not hold that *contact* was necessary.   I
farther stated the fact, that the experiment was a failure, as the patient
became affected, *not* under the *mes*merized, but under the *un*mesmerized
trees, which led the Commission to infer, that the phenomena resulted
from imagination, and not from the influence of a magnetic fluid.   Here,
then, we had two theories, neither of which considered contact necessary.
Surely no one could suppose that I wished to lay claim to these facts as
discoveries of my own, seeing I gave the dates when the occurrence took
place, which was many years before I was born.
    Moreover, I explained, at the same lecture, the different modes of mes-
merizing, by passes *at a distance*, and by pointing the fingers at the eyes

tion of the nervous system induced by both modes to be at least analogous. It appeared to me that the fixation of the mind and eyes was attained occasionally during the monotonous movements of the mesmerizers, and thus they succeeded sometimes, and as it were, by chance; whereas, by my insisting on the eyes being fixed in the most favourable position, and the mind thus riveted to one idea, as the *primary and imperative conditions*, my success was consequently general and the effects intense, while theirs was feeble and uncertain. However, from what the mesmerizers state as to effects which they can produce in certain cases, there seem to be differences sufficient to warrant the conclusion that they ought to be considered as distinct agencies ; and for the following reasons. The mesmerizers positively assert that they can produce certain effects, which I have never been able to produce by my mode, although I have tried to do so.[1] Now, I do not consider it fair or proper to impugn the statements of others in this matter, who are known to be men of talent and obser-

and forehead, adopted by others, long before I made any experiments on the subject ; and at subsequent lectures, from observing the graceful attitudes some patients assumed during the hypnotic state, and the ease with which they could maintain any given position, by becoming cataleptiformly fixed in it, I hazarded the opinion, that it may have been to hypnotism the Grecians were indebted for their fine statuary; and the Fakirs for their power of performing their remarkable feats. I also expressed my belief, that the rapt state of religious enthusiasts, such as that of the monks of Mount Athos, arose from the same cause, although none of the parties might have understood the true principle by which they were produced.

[1] The effects I allude to are such as, telling the time on a watch held behind the head, or placed on the pit of the stomach ; reading closed letters, or a shut book; perceiving what is doing miles off; having the power of perceiving the nature and cure of the diseases of others, although uneducated in medical science ; mesmerising patients at miles' distance, without the knowledge or belief in the patient that any such operation is intended.

vation, and of undoubted credit in *other* matters, merely because *I* have not *personally* witnessed the phenomena, or been able to produce them myself, either by my own mode or theirs.   With my present means of knowledge I am willing to admit that certain phenomena to which I refer *have* been induced by others, but still I think most of them may be explained in a different and more natural way than that of the mesmerizers.   When I shall have personally had evidence of the special influence and its effects to which they lay claim, I shall not be backward in bearing testimony to the fact.

   However, the greatest and most important difference is this, that they can succeed so seldom, and I so generally, in inducing the phenomena which we both profess thus to effect.   Granting, therefore, to the mesmerizers the full credit of being able to produce certain wonderful phenomena which I have not been able to produce by my plan, still it follows, that mine is superior to theirs in as far as *general applicability and practical utility are concerned.*   Mine has also this advantage, that I am quite certain no one can be affected by it, in any stage of the process, unless by the free will and consent of the patient, which is at once sufficient to exonerate the practice from the imputations of being capable of being converted to immoral purposes, which has been so much insisted on to the prejudice of animal magnetism   This has arisen from the mesmerizers asserting that they have the power of overmastering patients irresistibly, even whilst at a distance, by mere volitions and secret passes.

   I am fully borne out by the opinion of that eminent physiologist, Mr Herbert Mayo, in my view of the subject, that my plan is "the best, the shortest, and surest for getting the sleep," and throwing the nervous system, by artificial contrivance, into a new condition, which may be rendered available in the healing art. At a private conversazione, which I gave to the profession in London on the 1st of March, 1842, he examined and tested my patients most carefully, submitted himself to be operated on by me both publicly and privately, and was so searching and inquisitive in his investigations as to call forth the animadversions of a medical gentleman present, who thought he was not giving me fair play; but which he has assured me proceeded from an anxious desire to know the truth, not being biassed by having any peculiar views of his own to bring forward; and because he considered the subject most important, both in a speculative and practical point of view.

Whatever I advance, therefore, in the following remarks, I wish to be distinctly understood as strictly in reference to my own mode of operating, and distinct from that of all others. The latter I shall merely refer to in as far as is necessary to point out certain sources of fallacy by which the phenomena of the one may be confounded with those of the other.

In proof of the general success of my mode of operating, I need only name, that at one of my public lectures in Manchester, fourteen male adults, in good health, all strangers to me, stood up at once, and ten of them became decidedly hypnotized. At Rochdale I conducted the experiments for a friend, and hypnotized twenty strangers in one night. At a private conversazione to the profession in London, on the 1st of March, 1842, eighteen adults, most of them entire strangers to me,

sat down at once, and in ten minutes sixteen of them
were decidedly hypnotized.   Mr Herbert Mayo tested
some of these patients, and satisfied himself of the
reality of the phenomena.   On another occasion I took
thirty-two children into a room, none of whom had
either seen or heard of hypnotism or mesmerism : I
made them stand up at three times, and in ten or
twelve minutes had the whole thirty-two hypnotized,
maintaining their arms extended while in the hypnotic
condition, and this at mid-day.   In making this statement, I do not mean to say they were in the *ulterior*
stage, or state of *torpor;* but that they were in the *primary*
stage, or that of *excitement*, from which experience has
taught me confidently to rely that the torpid and rigid
state will certainly follow, by merely affording time for
the phenomena to develop themselves.   At the conversazione given on the 29th of June, 1842, to the members
of the British Association, two men and two youths
were brought off the street.   One man and both youths
were operated on ; all the three were hypnotized, and
one of the youths reduced to the rigid state.   In the
Stockport Chronicle of 4th February 1842, there is a
report of a lecture delivered in that town a few days
before.   A dozen male patients were made to stand up
at once, and treated according to my method, six of
them became hypnotized, and two of them so deeply,
as to cause the lecturer very considerable trouble to
rouse them.   With one named " Charlie," all the usual
means, including buffetings and frictions before a fire,
did not succeed in restoring speech until he had been
made to swallow nearly half a tumbler glass of *neat gin*.
I consider this important as being the testimony of *an
enemy*.   It can take place also in the dark, as well as by

day or by gas light; when the eyes are bandaged, as
when they are uncovered, by merely keeping the eyes
fixed, the body in a state of absolute rest, and the mind
abstracted from all other considerations.   In cases of
children, and those of weak intellect, or of restless and
excitable minds, whom I could not manage so as to
make them comply with these simple rules, I have
always been foiled, although most anxious to succeed.
This I consider a strong proof of the correctness of my
views.   By arresting the attention, and fixing the eyes,
it is also successful with brute animals.

This general success of my plan, both with man and
brute animals, I consider sufficient to prove it proceeds
from a law in the animal economy.   The exceptions to
success are so few as to lead to the conclusion that they
arise from a non-compliance with the conditions.   It is,
however, unquestionable, that there exists great difference in the susceptibility of different individuals, some
becoming rapidly and intensely affected, others slowly
and feebly so.

I am aware that some say they have tried my mode,
and failed to produce the phenomena.   The reason, I
presume, is simply this.   They will not believe the
necessity of complying with the WHOLE of the conditions I have distinctly insisted on.   But, in all fair-
ness, if they do not comply with the WHOLE conditions,
they have no right to expect the promised results, nor
to be disappointed because they fail.   If the patient and
operator comply in *all* respects as I direct, success is
almost certain ; but, on the contrary, he is almost
equally certain to fail if *all* the conditions are not
strictly complied with.

When we consider the great difficulty to some persons

of abstracting their minds, and the greater difficulty of
ensuring that patients operated on in a public room
shall be able to abstract their minds entirely from the
circumstances with which they are surrounded, and from
other considerations concentrate their ideas entirely on
the subject in hand, and the equally great difficulty
of securing absolute quiet where a large number of
people are assembled, and the extreme quickness of
hearing when patients are passing into the hypnotic
state, which makes them liable to be roused by the
slightest noise, it must be evident, that a public lecture-
room is by no means a favourable place for operating
on patients for the first time.

Prosecuting the investigation, as I have been doing,
by experiments and observations, I have, as might be
expected, had occasion to modify and alter some of
my views and manipulations ; but still the principle
remains the same.

CHAPTER II

l NOW proceed to detail the mode which I practise for inducing the phenomena. Take any bright object (I generally use my lancet case) between the thumb and fore and middle fingers of the left hand ; hold it from about eight to fifteen inches from the eyes, at such position above the forehead as may be necessary to produce the greatest possible strain upon the eyes and eyelids, and enable the patient to maintain a steady fixed stare at the object.[1] The patient must be made to understand that he is to keep the eyes steadily fixed on the object, and the mind riveted on the idea of that one object. It will be observed, that owing to the consensual adjustment of the eyes, the pupils will be at first contracted : they will shortly begin to dilate, and after they have done so to a considerable extent, and have assumed a wavy motion, if the fore and middle fingers of the right hand, extended and a little separated, are carried from the object towards the eyes, most probably the eyelids will close involuntarily, with a vibratory

[1] At an early period of my investigations, I caused the patients to look at a cork bound on the forehead. This was a very efficient plan with those who had the power of converging the eyes so as to keep them *both* *steadily* directed on the object. I very soon found, however, that there were many who could not keep *both* eyes steadily fixed on *so near* an object, and that the result was, that such patients did not become hypnotized. To obviate this, I caused them to look at a more distant point, which, although scarcely so rapid and intense in its effects, succeeds more generally than the other, and is therefore what I now adopt and recommend.

motion.   If this is not the case, or the patient allows
the *eyeballs* to *move*, desire him to begin anew, giving
him to understand that he is to allow the eyelids to
close when the fingers are again carried towards the
eyes, but that the *eyeballs must be kept fixed in the same
position*, and the *mind riveted to the one idea of the object
held above the eyes.*   It will generally be found, that the
eyelids close with a *vibratory* motion, or become spasmodically closed.   After ten or fifteen seconds have
elapsed, by gently elevating the arms and legs, it will
be found that the patient has a disposition to retain
them in the situation in which they have been placed,
if *he is intensely affected.*   If this is not the case, in
a soft tone of voice desire him to retain the limbs in
the extended position, and thus the pulse will speedily
become greatly accelerated, and the limbs, in process
of time, will become quite rigid and involuntarily fixed.
It will also be found, that all the organs of special sense,
excepting sight, including heat and cold, and muscular
motion, or resistance, and certain mental faculties, are
at *first* prodigiously *exalted*, such as happens with regard
to the primary effects of opium, wine, and spirits.   After
a certain point, however, this exaltation of function is
followed by a state of depression, far greater than the
torpor of *natural* sleep.[1]   From the state of the most
profound torpor of the organs of special sense, and tonic
rigidity of the muscles, they may, at this stage, *instantly*

---

[1] I wish to direct especial attention to this circumstance, as from overlooking the fact of the *first* stage of this artificial hypnotism being one of
excitement, with the possession of *consciousness* and *docility*, many imagine
they are *not* affected, whilst the acceleration of pulse, peculiar expression
of countenance, and other characteristic symptoms, prove the existence of
the condition beyond the possibility of a doubt, *to all who understand the
subject.*   I consider it very imprudent to carry it to the ulterior stage, or

be restored to the *opposite* condition of extreme mobility and exalted sensibility, by directing a current of air against the organ or organs we wish to excite to action, or the muscles we wish to render limber, and which had been in the cataleptiform state. By mere repose the senses will speedily merge into the original condition again. The *modus operandi* of the current of air producing such extraordinary effects, I acknowledge myself quite unable to explain, but I have no difficulty in producing and reproducing the effects by the same means, whether performed by myself or others, and whether the current of air is from the lips, from a pair of bellows, or by the motion of the hand, or any inanimate object. The extent and abruptness of these transitions (see page 136) are so extraordinary, that they must be seen before the possibility is believed.

An abrupt blow, or pressure over the rigid muscle, will de-hypnotize a rigid part ; but, I have found pressing the nose will not restore smell, unless very gentle and continued, nor will pressing a handkerchief against the ear restore hearing when the ear has become torpid, nor will *gentle* friction over the skin restore sensibility to the dormant skin, or mobility to the rigid muscles underneath (unless so gentle as to be titillation, properly so called), and yet a slight puff of wind will *instantly* rouse the whole to abnormal sensibility and mobility : a fact which has perplexed and puzzled me exceedingly.

that of torpor, at a *first* trial.   Moreover, there is great difference in the susceptibility to the neuro-hypnotic impression, some arriving at the state of rigidity and insensibility in a few minutes, whilst others may readily pass into the *primary* stage, but can scarcely be brought into the ulterior, or rigid and torpid state.   It is also most important to note, that many instances of remarkable and permanent cures have occurred, where it has never been carried beyond the state of consciousness.

At first I required the patients to look at an object until the eyelids closed of themselves, involuntarily. I found, however, that in many cases this was followed by pain in the globes of the eyes, and slight inflammation of the conjunctival membrane. In order to avoid this, I now close the eyelids, when the impression on the pupil already referred to has taken place, because I find that the *beneficial* phenomena follow this method, provided the eyeballs are kept fixed, and thus, too, the unpleasant feelings in the globes of the eyes will be prevented. Were the object to produce astonishment in the person operated on, by finding himself unable to open his eyes, the former method is the better ; as the eyes once closed it is generally impossible for him to open them ; whereas they may be opened for a considerable time after being closed in the other mode I now recommend. However, for curative purposes, I prefer the plan which leaves no pain in the globes of the eyes.

In fine, from a careful analysis of the whole of my experiments, which have been very numerous, I have been led to the following conclusion :—That it is a law in the animal economy, that by a continued fixation of the mental and visual eye, on any object which is not of itself of an exciting nature, with absolute repose of body, and general quietude, they become wearied ; and, provided the patients rather favour than resist the feeling of stupor of which they will soon experience the tendency to creep upon them, during such experiments, a state of somnolency is induced, accompanied with that condition of the brain and nervous system generally, which renders the patient liable to be affected, according to the mode of manipulating, so as to exhibit the hypnotic phenomena. As the experiment succeeds with the blind, I consider it not so much the optic, as the sentient, motor, and sympathetic nerves, and the mind through which the impression is made. I feel so thoroughly convinced that it is a law of the animal economy that such effects should follow such condition of mind and body, that I hesitated not to give it as my deliberate opinion, that this is *a fact* which cannot be controverted. As to, the *modus operandi* we may never be able to account for that in a manner so as to satisfy all objections; but neither can we tell why the law of gravitation should act as experience has taught us it *does* act. Still, as our ignorance of the cause of gravitation acting as it is known to do, does not prevent us profiting by an accumulation of the facts known as to its results, so ought not our ignorance of the *whole* laws of the hypnotic state to prevent our studying it practically, and applying it beneficially, when we have the power of doing so.

I feel confident that the phenomena are induced solely by an impression made on the nervous centres, by the physical and psychical condition of the patient, irrespective of any agency proceeding from, or excited into action by another—as any one can hypnotize himself by attending strictly to the simple rules I lay down; and the following is a striking example of the fact, which was communicated to me and two other gentlemen, by a most respectable teacher. He found that a number of his pupils had been in the habit of hypnotizing themselves, and he had ordered them to discontinue the practice. However, one day he ascertained a girl had hypnotized herself by looking at the wall, and that her companions had put a pen in her

H

hand, with which she had written the word "Manchester;" and she held the pen very firmly—in fact the fingers were cataleptiformly rigid. He spoke to her in a gentle tone of voice, and called her. She arose and advanced towards him, and when awoke, was not aware he had called her, or of what had passed.

I have also had the state of the patient tested before, during, and after being hypnotized, to ascertain if there was any alteration in the magnetic or electric condition, but although tested by excellent instruments, and with great care, no appreciable difference could be detected. Patients have been hypnotized whilst positively, and also whilst negatively, electrified, without any appreciable difference in the phenomena; so that they appear to be excited independently of electric or magnetic change. I have also repeatedly made two patients hypnotize each other, at the same time, by personal contact. How could this be reconciled with the theory of a special influence transmitted being the cause of the phenomena, *plus* and *minus* being equally efficient?

It is also well known, that occasionally the phenomena arise spontaneously in the course of disease.

It is now admitted even by the editor of the *Lancet*, one of the greatest opponents of mesmerism, in the leading article of 4th February, 1843, that the phenomena "are wonderful only to those who are unacquainted with the aspects of disease;" and "that we continually see patients labouring under hysteria, and analogous forms of nervous disease, falling suddenly into various states of stupor, trance, and convulsion, without *any* assignable cause." When it is acknowledged that such effects as those named, may spring from such slight influences as to be said to arise "*without any assignable cause,*" can it be wondered at that important changes may be induced by acting on the nervous system in the way I have adopted, of which Mr Herbert Mayo (whose competence to give an opinion on *any* physiological subject no one will question, and who himself publicly submitted to be operated on by me) observed, in the course of our correspondence, that it induces "a feeling of stupor, which any one may observe has a disposition to creep upon him, when he tries your experiment of looking fixedly at an object as you direct."

I thought it desirable, therefore, to adopt the name I did, for the reasons explained in the introduction.

A patient may be hypnotized by keeping the eyes fixed in *any* direction. It occurs most *slowly* and *feebly* when the eyes are directed straight forward, and most *rapidly* and *intensely* when they can be maintained in the position of a double internal and upward squint.[1]

---

[1] It is not a little amusing to find any one try to distort so greatly, by garbled statements, the plain meaning of an author, as to make it appear that a writer of some articles on Animal Magnetism, in the *Medical Gazette* in 1838, was well acquainted with my mode of operating. He observes at page 856, "On the majority of persons no influence whatever is exhibited." How does this coincide with the general success of my mode as stated at page 107 ? "On those least affected a number of anomalous slight symptoms are produced." He then describes those "feelings of heat and cold, and those of creeping and trembling," which, he adds, "are only the usual imaginary feelings which most persons have if their attention be strongly directed to any particular part of the body, more especially if (as is generally the case with magnetic patients) something is expected to occur." Such are the symptoms attributed by this writer to "attention," but are these the symptoms or phenomena induced by Hypnotism, as stated in Chapter IV. ? Or is there the slightest similarity in the cause? In this author's view it is the result of "attention strongly directed *to different parts of the body,*" whereas mine is by attention riveted to something *without* the body. The best mode of gathering the opinion of an author appears to me to be that of his summing up at the conclusion of his subject. Now, at page 1037, the subject is concluded by the following observations :—"This, then, is our case. Every credible effect of magnetism has occurred, and every incredible *is said* to have occurred in cases

It is now pretty generally known, that during the effort to look at a very near object, there is produced, according to the direction of the object, a double internal squint, or double internal and downward or upward squint, and the pupils are thereby powerfully contracted. I am not aware, however, that it has been recorded, that by directing the eyes loosely, upwards or downwards, to the right or to the left, as if looking at a very distant object, the pupils become very much *dilated*, irrespective of the quantity of light passing to the retina ; so that in this manner we can contract or dilate the pupil at will. To those who consider the movement of the iris as the mere effect of irritability, I may observe, in that view, the former position increases, the latter diminishes, the irritability. I may farther remark, if the eyes are *much strained* in ANY direction, I think the pupils will be found to contract as a consequence.

It is important to remark, that the oftener patients are hypnotized, from association of ideas and habit, the more susceptible they become ; and in this way they are liable to be affected *entirely through the imagination.*

where no magnetic influence has been exerted, but in all which excited imagination, irritation, or some powerful mental impression, has operated : where the mind has been alone acted on, magnetic effects have been produced without magnetic manipulations : where magnetic manipulations have been employed, unknown, and therefore without the assistance of the mind, no result has ever been produced." Now, can any thing more be required than this, to prove that this writer, as well as Bertrand, adheres to the theory of imagination? Such was the impression left on my mind by reading these papers when they were published ; and, together with Wakley's experiments, determined me to consider the whole as a system of collusion or illusion, or of excited imagination, sympathy, or imitation. I therefore abandoned the subject as unworthy of farther investigation, until I attended the conversazioni of Lafontaine, where I saw one fact, the inability of a patient to open his eyelids, which arrested my attention ; I felt convinced it was not to be attributed to any of the causes referred to, and I therefore instituted experiments to determine the question ; and exhibited the results to the public in a few days after.

Thus, if they consider or imagine there is something doing, although they do not see it, from which they are to be affected, they *will become affected;* but, on the contrary, the most expert hypnotist in the world may exert all his endeavours in vain, if the party does not expect it, and mentally and bodily comply, and thus yield to it.

It is this very circumstance, coupled with the extreme docility and mobility of the patients, and extended range and extreme quickness of action, at a certain stage, of the ordinary functions of the organs of sense, including heat and cold, and muscular motion, the tendency of the patients in this state to approach to, or recede from, impressions, according as their intensity or quality is agreeable or the contrary, which I consider has misled so many, and induced the animal magnetizers to imagine they could produce their effects on patients at a distance, through mere volition and secret passes.[1]

[1] In the *Medical Times* of 26th March, 1842, I published a letter on this subject, from which I make the following extracts :—

" The supposed power of seeing with other parts of the body than the eyes, I consider is a misnomer, so far as I have yet personally witnessed. It is quite certain, however, that some patients can tell the shape of what is held at an inch and a half from the skin, on the back of the neck, crown of the head, arm, or hand, or other parts of the body, but it is from *feeling* they do so ; the extremely exalted sensibility of the skin enabling them to discern the shape of the object so presented, from its tendency to emit or absorb caloric. This, however, is not *sight*, but *feeling*.

" In like manner I have satisfied myself and others, that patients are drawn, or induced to obey the motions of the operator, not from any peculiar inherent magnetic power in him, but from their exalted state of feeling enabling them to discern the currents of air, which they advance to, or retire from, according to their direction. This I clearly proved to be the case to-day, and that a patient could feel and obey the motion of a glass funnel passed through the air at a distance of *fifteen feet*.

" To remove all sources of fallacy as to the extent of influence exercised by the patient herself, independently of any personal or mental influence on my part, whilst I was otherwise engaged, my daughter requested the patient to go into a room by herself, and, when alone, try whether she

It would be difficult to adduce a more striking example than the following of the fact, that the phenomena are produced by the fixation of the mind and eyes, and general repose of the patient, and not from imagination, or the look or will of another. After my lecture at the Hanover Square Rooms, London, on the 1st of March, 1842, a gentleman told Mr Walker, who was along with me, that he was most anxious to see me, that I might try whether I could hypnotize him. He said both himself and friends were anxious he should be affected, but that neither Lafontaine nor others who had tried him, could succeed. Mr Walker said, if that is what you want, as Mr Braid is engaged otherwise, sit down, and I will hypnotize you myself in a minute. When I went into the room I observed what was going on, the gentleman sitting staring at Mr Walker's finger,

could hypnotize herself. In a short time I was told the patient was found fast asleep in my drawing-room. I went to her, bandaged her eyes, and then, with the glass funnel (which I used to avoid the chance of electric or magnetic influence being passed from my person to that of the patient), elevated, or drew up her arms, and then her whole body. I now retired fifteen feet from her, and found every time I drew the funnel *towards* me, she approached nearer, but when it was forced sharply *from me*, she invariably retired ; and if it was moved laterally, she moved to the right or left accordingly. I now continued drawing the funnel so as to keep up the currents towards the door, and in this way, her arms being extended, and eyes bandaged, she followed me down stairs and up again, a flight of twenty-two steps, with the peculiar characteristic caution of the somnambulist. After arriving at the top of the stair, I allowed her to stand a little, and again began the drawing motion. She evidently felt the motion, and attempted to come, but could not. I now endeavoured to lead her by the hand, but found that *the legs had become cataleptiform, so that she could not move.* I now carried her into the drawing-room, and, after she was seated on a chair, awoke her. She was quite unconscious of what had happened, and could not be made to believe she had been down stairs —said she was quite sure she could have done no such thing without falling—and to this moment believes we were only hoaxing her by saying she had had such a ramble.

"I had repeatedly performed this experiment with this patient and who was standing a little to the right of the patient, with his eyes fixed steadily on those of the latter.   I passed on, and attended to something else, and when I returned a little after, found Mr Walker standing in the same position *fast asleep, his arm and finger in a state of cata-leptiform rigidity*, and the patient wide awake, and staring at the finger all the while.   After I had roused Mr Walker, the gentleman observed, " this is really very strange, that no one can mesmerise *me ;* I must have extraordinary powers of resistance."   I requested him to stay a little, and I would try what I could do for him when all was quiet.   In three minutes I had him asleep, and in a little more quite rigid.   The following reasons may be assigned for my success after Mr Walker had so signally failed.   He tried it whilst there were several people in the room, who were moving about and talking ;

others before, with the same result in all respects but walking up and down stairs ; and proved their readiness to be drawn by others equally as myself when in that state ; so that I consider it quite evident to any un-prejudiced person, that a patient can hypnotize himself independently of any personal influence of another ; and that it is by extreme sensibility of the skin, and docility of the patients, that they are drawn after an operator, rather than by magnetic attraction ; and that the power of discriminating objects held near the skin in different parts of the body, is the result of *feeling*, and *not of sight.*

" The moment I witnessed the attempts of a celebrated professor *to draw a patient*, I formed my opinion of the *cause ;*—that it arose from currents of air produced by his hand, together with the extreme sensibility of the skin, and docility of the patients when in that state ; and my experiments have clearly proved this, *some patients acknowledging the fact.*

" It may be interesting to remark, that whilst passing up and down stairs the door bell rang, which produced such a tremor through the whole frame as nearly caused the patient's fall—a fact quite in accordance with the effect of any abrupt noise on NATURAL *somnambulists.*"

It is owing to this extreme sensibility of the skin during hypnotism, that patients may walk through a room blindfolded, without running against the furniture—the difference of temperature, or rather degree of conduct-ing power of objects, and the resistance of the air directing them.

I have frequently illustrated this with very sensitive patients in the most

I took care not to commence till all was quiet—Mr
Walker had not taken the precaution to make the
patient direct his eyes in the best possible manner, but
I was careful that he should do so.   Moreover, although
Mr Walker had not succeeded in putting him into the
somnolent condition, he had, no doubt, partially affected
him, and the influence had not entirely passed off when
I began my operation.   Two days after, Mr Walker
accompanied me when I called on one of the most celebrated mesmerizers in Europe, who, during our conversation, stated, that a glance of the eye was quite enough,
in many cases, to produce the effects.   During our
conversation, I presume, he had determined to surprise
both Mr Walker and myself, by keeping his large

beautiful and satisfactory manner, thus : By throwing any fragrant and
agreeable scent on a bare table the patients will approach, anxious to smell
it, but are repelled before they come quite close to the cold table.   Place
a handkerchief on the table, on which place the scent, and now the patient
will approach close to it, and revel in its fragrance.   Remove the handkerchief, and the attractive and repulsive movements will again ensue.

This was beautifully illustrated at a private conversazione at my house
lately, in the presence of several medical and other eminently scientific
gentlemen.   Two patients were hypnotized, when one became so enam-
oured of the scent of a gentleman's snuff-box as to follow him round the
room.   He then laid the box about eighteen inches from the edge of an
uncovered table, when she advanced, her arms being extended, anxious to
reach the box, but when about ten or twelve inches from it, she started
back, from perceiving the impression of the cold table at that distance.
She now made another attempt to approach the box, being attracted by
the fragrance of its contents, but was as speedily repelled by the cold
table before she approached it, and now kept bobbing over the box, much
in the same manner as I have witnessed in the attempts of a hungry dog
to partake of very hot food.   The other patient, in passing round the
table, also caught the smell of the box, and advanced from another point,
and thus both kept bobbing over it, much to the amusement of all present.
I now covered the table with a handkerchief, and placed the box on it,
when they instantly approached close to it, and seemed to feast on its
fragrance ; on removing the handkerchief they withdrew, and commenced
bobbing over it as at first.   The former patient had never seen such
experiments, or been tested in this way before.

intellectual eyes fixed on Mr Walker.   The latter, however, suspecting what was intended, and knowing my
opinion as to the mode of *resisting* the influence of *such
fascination,* kept his eyes moving, and his mind roaming,
and thus frustrated the volition of one of the most energetic minds, and the glances and fascination of one of
the finest pair of eyes imaginable for such a purpose.
I must remark, that Mr Walker was once magnetized
by M. Lafontaine, after having been several times
operated on by me, a circumstance which of course
would render him more susceptible to the influence of
the animal magnetizers' modes of operating, according
to their own theory.   Had Mr Walker believed in the
power, I know he would have become affected, even
supposing the gentleman referred to had no such intention,—and I am not prepared to say he had.   Mr
Walker, however, firmly believed he was trying to
mesmerize him by the fascination referred to ; but, re-
lying on my opinion, and acting accordingly, he escaped.
In order to show the efficacy of my simple plan, in a
short time after, in the presence of the same gentleman,
I requested Mr Walker to hypnotize himself.   By
simply fixing his eyes and mind this was accomplished
in about a minute.

CHAPTER III

I CONSIDER it unnecessary, in this treatise, to enter into a *detailed* account of the ordinary phenomena of sleep, dreaming and somnambulism, as contrasted with the waking state. Suffice it to say, the waking condition is that of mental and bodily activity, during which we are enabled to hold communion with the external world, by perceiving the ordinary impressions of appropriate stimuli through the organs of special sense, and of exercising the power of voluntary motion, and the mental functions generally. The state of *profound* sleep is exactly the *reverse* of this—a state of absolute *un*consciousness of all that is going on around, and suspension of voluntary motion, and intellectual activity. In as far as regards the organs of special sense, and voluntary motion, and a temporary suspension of the mental energies, it is the emblem of death.

Between these extreme points there are *gradual* transitions, so that there are all possible varieties of condition imaginable, from the highest state of mental and bodily activity, to absolute torpor of both. There are two conditions, however, to which I may *briefly* advert,—that of dreaming and of somnambulism. In the former there are some of the mental and bodily functions in a state of partial activity, but, from the sensations arising from external stimuli being perceived very imperfectly, erroneous impressions are conveyed to the mind; and, as

happens in some cases of insanity, the power of controlling the current of thought being absent, one idea excites
another, until the most incongruous combinations are
produced in many instances. Somnambulism, properly
so called, is a state still more nearly allied to the waking
condition than dreaming. The mental functions are
more awake, a more just estimate of external impressions can be formed, and there is the power of voluntary
motion present in a remarkable degree. Persons in this
state are thus capable of being directed by those around,
into certain trains of thought and action. The principal
difference between the natural somnambulists, and those
who become so through hypnotizing in the manner
pointed out in this treatise, is the greater tendency of
the latter to lapse into a state of *profound* sleep, unless
prevented by being roused and directed by those present. Natural somnambulists seem to be impelled to
certain trains of action by *internal* impulses ; but, so far
as I have seen, the artificial somnambulists have an inclination to remain at absolute rest, unless excited to
action by some impression from without. In compliance with such excitement, however, they evince great
acuteness and docility. There is also another remarkable difference. It is stated, that although natural somnambulists cannot remember, when awake, what they
were engaged in when asleep, they have a vivid recollection of it when in that state again ; but I have found
no parallel to *this* in the somnambulism induced by
hypnotism.

By this I mean that they cannot explain what happened during the former somnambulic state, but they
may approximate to the words and actions which had
formerly manifested themselves, provided they are

placed under exactly similar circumstances. For the extent to which peculiar manifestations may be brought out by manipulating the head and face, at a certain stage of hypnotism, see Chapter VI. where examples are given of memory as regarded events which happened during the *waking* condition, whilst they seemed to have no recollection of what happened during a former state of hypnotism.

As to the causes of common sleep, I may remark, that, by the exercise of the mental operations, and the impressions conveyed through the organs of special sense, muscular effort, and the discharge of other animal functions, the brain becomes exhausted, and ceases to be affected by ordinary stimuli, and lapses into that dormant state we call sleep. During this condition it becomes recruited, and fitted for again receiving its wonted impressions through the organs of sense, and of holding intercourse with external nature, and exercising those powers of voluntary motion and mental function peculiar to the waking condition.

It will be generally admitted, that the most refreshing, and therefore the *most natural* sleep, accompanies that condition or languor which follows the *moderate* exercise or fatigue of *all* the bodily and mental functions, rather than an undue exercise of *one* or *more* to the neglect of the others. It is long since it was observed that inordinate attention to one subject caused *dreaming* instead of *sound sleep*. It will also be found that the absolute length of time during which any function may be exercised, depends very much on the *continuity* of its exertion, or its alternation with that of other functions ; thus the mind may become confused and bewildered by continuing one particular study for a length of time, but may be able to return to it with energy and advantage, and prosecute the subject longer on the whole, by varying it with study of a different nature ; moreover, bodily disease, and even insanity, frequently arises from allowing the mind to be occupied inordinately by one particular object or pursuit, whether that may be religion, politics, avarice, schemes of ambition, or any other passion, emotion, or object of unvaried contemplation.

In like manner, continued and over-intense muscular effort very soon exhausts the power of the muscles so exercised or over-exerted ; and by keeping the eyes steadily and constantly exercised by gazing on a coloured spot, they soon cease to be able to discern the boundaries of the respective colours,[1] and ultimately seem scarcely to be capable of distinguishing the spot at all. The same might be proved of the other senses. In fine, *alternate action and repose is the law of animated nature.*[2]

[1] Müller.

[2] This subject is beautifully illustrated by Müller, at page 1410, Vol. II. (Baly's translation), which I now quote. "The excitement of the organic processes in the brain which attends an active state of the mind, gradually renders that organ incapable of maintaining the mental action, and thus induces sleep, which is to the brain what bodily fatigue is to other parts of the nervous system. The cessation or remission of mental activity during sleep, in its turn, however, affords an opportunity for the restoration of integrity to the organic conditions of the cerebrum, by which they regain their excitability. The brain, whose action is essential to the manifestation of mind, obeys, in fact, the general law which prevails over all organic phenomena,—that the phenomena of life being particular states induced in the organic structures, are attended with changes in the constituent matter of these structures. Hence, the longer the action of the mind is continued, the more incapable does the brain become of supporting that action, and the more imperfectly are the mental processes performed, until at length sensations cease to be perceived, notwithstanding the impressions of external stimuli continue. This is entirely analogous to what frequently occurs during the waking state, in the case of individual sensations."

I must beg leave to take one exception to the correctness of these remarks, and that is, *moderate* exercise, I consider, instead of *exhausting,*

It is this very principle, of over-exerting the attention, by *keeping it riveted to one subject or idea which is not of itself of an exciting nature*, and over-exercising one set of muscles, and the state of the strained eyes, with the suppressed respiration, and general repose, which attend such experiments, which excites in the brain and whole nervous system that peculiar state which I call Hypnotism, or nervous sleep. The most striking proofs that it is different from common sleep, are the extraordinary effects produced by it.[1]   In deep

seems rather to act *as a salutary stimulus*, and thus *strengthens* both *organ and function.*   He then goes on to state, most lucidly and fairly, " Nor merely the action of the mind, but the long continued exertion of other functions of animal life, such as the senses or muscular actions, induces the same exhaustion of the organic states of the brain, and thereby want of sleep and sleep itself ; for these different systems of the body participate in the change which the organic condition of any one of them may undergo. Lastly, impairment of the normal organic state of the brain, by the circulation through it of blood charged with imperfectly assimilated nutriment, as after full meals in which spirituous drinks have been taken, also induces sleep.   The narcotic medicaments act still more strongly by the change they produce in the organic composition of the sensorium.   Even the increased pressure of the blood upon the brain, produced by the horizontal posture, may become the cause of sleep."

Here then is the opinion of this author in a few words.   The exercise of function is attended with a change, deterioration, or wasting of the organic structure at a more rapid rate than can be repaired by the slow, but regular and persistent organic renovation continually going on in the whole system.   A cessation of sentient, and mental, and muscular functions, therefore, as happens in sleep, becomes necessary to afford time for the renovation of the deteriorated organic structures of the respective organs, and of the brain in particular, which, in so eminent a degree, sympathizes and participates in the organic changes which have been induced in other organs.

Liebig's views seem confirmatory of this, where he points out the fact, that the chemical principles of those substances which act most energetically on the brain and nerves have a composition analogous to that of the substance of the brain and nerves, as in the case of the vegetable alkaloids. He believes that all the active principles which produce powerfully poisonous or medicinal effects, in minute doses, are compounds of nitrogen ; and that those compounds, being resolved into their elements, take a share in the formation, or transformation, of brain and nervous matter.

[1] See Appendix I., Note 7.

abstraction of mind, it is well known, the individual becomes unconscious of surrounding objects, and in some cases, even of severe bodily inflictions. During hypnotism, or nervous sleep, the functions in action seem to be so *intensely* active, as must in a great measure rob the others of that degree of nervous energy necessary for exciting their sensibility. This alone may account for much of the dulness of common feeling during the abnormal quickness and extended range of action of certain other functions.[1]

[1] It was certainly presuming very much on the ignorance of others for any one to attempt so to pervert the meaning of an author, as to twist what M'Nish has written on the article "Reverie," and represent it as the basis of my theory. How does M'Nish define it? "Reverie," he says, "proceeds from an unusual quiescence of the brain, and inability of the mind to direct itself strongly to any one point; it is often the prelude of sleep. There is a *defect* in the *attention*, which, instead of being fixed on *one* subject, *wanders over a thousand*, and even on these is feebly and ineffectively directed." Now this, as every one must own, is the very *reverse* of what is induced by *my plan*, because I *rivet* the *attention* to *one idea*, and the eyes to *one point*, as the *primary* and *imperative conditions*. Then, as to another passage, "That kind of reverie in which the mind is nearly divested of all ideas, and approximates nearly to the state of sleep, I have sometimes experienced while gazing long and intently upon a river. The thoughts seem to glide away, one by one, upon the surface of the stream, till the mind is emptied of them altogether. In this state we see the glassy volume of the water moving past us, and hear its murmur, but lose all power of fixing our attention definitively upon any subject; and either fall asleep, or are aroused by some spontaneous reaction of the mind, or by some appeal to the senses sufficiently strong to startle us from our reverie." Now, I should have read this passage a thousand times without discovering any analogy between it and my theoretical views. They appear to me to be "wide as the poles asunder." Instead of ridding the mind of ideas "one by one, till the mind is *emptied* of them *altogether*," I endeavour to rid the mind at *once* of all ideas *but one*, and to fix *that* one in the mind *even after passing into the hypnotic state*. This is very different from what happens in the reverie referred to, in which M'Nish confesses the difficulty "of fixing our attention definitively upon *any* subject." Again, so far from a reaction of the mind being sufficient to rouse patients from the hypnotic state, as in the reverie referred to, I can only state, that I have never seen patients deeply affected come out of it without assistance; and I heard Lafontaine say,

The untoward result referred to in the note below, I have no doubt, was the effect of permitting the experiment to be carried too far. No such consequence has ever followed in any of my operations, and for this reason, that I have always watched each case with close attention, and aroused the patient the moment I saw the slightest symptom of danger. I shall, therefore, now point out the symptoms of danger, with the mode of arousing patients, and thus preventing mischief which might ensue from want of due caution in the operator.

Whenever I observe the breathing very much oppressed, the face greatly flushed, the rigidity excessive, or the action of the heart very quick and tumultuous, I instantly arouse the patient, which I have always readily and speedily succeeded in doing by a clap of the hands, an abrupt shock on the arm or leg by striking them sharply with the flat hand, pressure and

he had been unable to restore the Frenchman who was with him for twelve hours on one occasion, when a surgeon operated on him ; and I have read the report of another, who operated on a patient at Stockport, "Charlie," according to my method, and, from having allowed him to go too far, experienced no small difficulty in rousing him, nor could he be restored to speech after much manipulation, and buffeting, and friction, till he had swallowed nearly *half a tumbler glass of neat gin*. To prevent misrepresentation, I shall quote the case as reported in the *Stockport Chronicle* of 4th February, 1842.—"To the final instance the lecturer now drew particular attention. It was that of a young man, recognized by many in the room by the familiar name of 'Charlie.' He was just entering upon the state of somnolence, and the attention of the audience was directed to the fact, that it was so indicated, by the different members becoming rigid. Presently his eyelids closed, and he became as though apparently under the influence of catalepsy. It was tried to make him sit down, but his whole frame was perfectly rigid, and that object could not therefore be accomplished. He was then laid on the floor, and the usual means, with cold water added, were employed in order to bring him to a state of consciousness. After a time these partially succeeded, his limbs became once more supple, and he was set in a chair, apparently conscious, though his eyelids were not yet open. He was several times requested to open them, and as often made the most friction over the eyelids, and by a current of air wafted against the face. I have never failed by these means to restore my patients very speedily.[1]

I feel convinced hypnotism is not only a valuable, but also a perfectly safe remedy for many complaints, if judiciously used ; still it ought not to be trifled with by ignorant persons for the mere sake of gratifying idle curiosity. In all cases of apoplectic tendency, or where there is aneurism, or serious organic disease of the heart, it ought not to be resorted to, excepting with the precaution, that it may be in the mode calculated to *depress* the force and frequency of the heart's action.

vigorous efforts to do so, but was unable ; at last they were opened, and it was discovered that the operation had so far influenced the entire functions of his body, that he had for a time lost the power of utterance, the muscles of the throat and tongue still remaining in a state of the most perfect rigidity. In this state, and being affected by a tremor which seized every part of his person, the patient was conducted into an ante-room, and placed before a fire, while the operator continued to rub the parts, in order to excite them to renewed action, and to restore animation. All this, however, had not the desired effect for some time, during which the patient evinced feelings of considerable surprise at his condition ; but nevertheless was exceedingly lively, and made several efforts to speak, but could not. At last half a tumbler glass of neat gin was brought, the greater portion of which he drank off, and this partially restored the power of utterance, for he was afterwards able to articulate a little, and asked, though only in a whisper, for his hat ; and also requested that some water might be mixed with the remaining portion of the gin. He complained also of a sense of excessive fulness of the stomach ; and said, in answer to inquiries, that although not feeling cold, he was yet unable to resist the tremor which had seized him."

Was not this a beautiful illustration of the facility with which patients might be roused from this condition "*by a reaction of the mind?*" Nor was this the only case that evening, in which great difficulty had been experienced in rousing patients from the hypnotic state.

[1] See Appendix I., Note 8.

I

CHAPTER IV

IN passing into common sleep objects are perceived more and more faintly, the eyelids close, and remain quiescent, and all the other organs of special sense become gradually blunted, and cease to convey their usual impressions to the brain, the limbs become flaccid from cessation of muscular tone and action, the pulse and respiration become slower, the pupils are turned upwards and inwards, and are *contracted*. (Müller.)

In the hypnotic state, induced with the view of exhibiting what I call the hypnotic phenomena, vision becomes more and more imperfect, the eyelids are closed, but have, for a considerable time *a vibratory motion* (in some few they are forcibly closed, as by spasm of the orbiculares); the organs of special sense, particularly of smell, touch, and hearing, heat and cold, and resistance, are greatly *exalted*, and afterwards become blunted, in a degree far beyond the torpor of natural sleep; the pupils are turned upwards and inwards, but, contrary to what happens in *natural* sleep, they are greatly *dilated*, and highly insensible to light; after a length of time the pupils become contracted, whilst the eyes are still insensible to light. The pulse and respiration are, at first, slower than is natural, but immediately on calling muscles into action, a tendency to cataleptiform rigidity is assumed, with rapid pulse, and oppressed and quick breathing. The limbs are thus maintained in a state of tonic *rigidity* for any length of time I have yet thought it prudent to try, instead of that state of flaccidity induced by common sleep; and the most remarkable circumstance is this, that there seems to be no corresponding state of muscular exhaustion from such action.[1]

In passing into natural sleep, any thing held in the hand is soon allowed to drop from our grasp, but, in the artificial sleep now referred to, it will be held more firmly than before falling asleep. This is *a very remarkable difference.*

The power of balancing themselves is so great that I have never seen one of these hypnotic somnambulists fall. The same is noted of natural somnambulists. This is a remarkable fact, and would appear to occur in this way, that they acquire the centre of gravity, as if by instinct, in the *most natural, and therefore, in the most graceful manner*, and if allowed to remain in this position, they speedily become cataleptiformly and immovably fixed. From observing these two facts, and the general tendency and taste for dancing displayed by most patients on hearing lively music during hypnotism, the peculiarly graceful and appropriate movement of many when thus excited, and the varied and elegant postures they may be made to assume by slight currents of air, and the faculty of retaining

---

[1] The average of a great number of experiments gives me the following results: The rise in the pulse from mere muscular effort, to enable patients to keep their legs and arms extended for five minutes, is about 20 per cent. When in the state of hypnotism it is upwards of 100 per cent. By arousing all the senses, and the head and neck, it will speedily fall to 40 per cent. (that is, twice what it was when so tested in the natural condition), and by rendering the whole muscles limber, whilst the patient is in the state of hypnotism the pulse very speedily falls to, or even below, the condition it was before the experiment.

any position with so much ease, I have hazarded the
opinion, that the Greeks may have been indebted to
hypnotism for the perfection of their sculpture, and the
Fakirs for their wonderful feats of suspending their
bodies by a leg or an arm.[1]

It thus clearly appears that it differs from common
sleep in many respects, that there is first a state of
excitement as with opium, and wine, and spirits, and
afterwards a state of corresponding deep depression or
torpor.

In the case of two patients, symptoms very much the
same as those produced in them by the laughing gas,
were produced twice on each patient, and the only time
I know of their having been hypnotized. One lost the
power of speech for two hours, as happened also after
the gas. Both these patients had hypnotized themselves. There is a remarkable difference between the
hypnotic condition, and that induced by the nitrous
oxide. In the latter there is great, almost irresistible

---

[1] It has been suggested to me, that it can scarcely be doubted that the
Bacchanalians, who had no feeling of wounds ("non sentit vulnera
Mœnas,"—*Ovid*), and whose condition was a stupor different from common sleep ("Exsomnis stupet Œvias,"—*Horace*), were in the hypnotic
condition or nervous sleep, and therein excited to dance by music; and
that, as uneducated maid-servants, when under the full influence of that
state of nerve, move with the grace and peculiar action of the most
accomplished dancers of pantomimic ballet, there is reason to believe,
not merely that the perfect grace exhibited in the attitudes represented in
ancient sculpture and painting, was derived from studying the Bacchan-
alian and other mystic dancers, but that the movements used by stage-
dancers, in our days, have been transmitted to us by continued imitation,
through Italy, from the dancers in the Greek mysteries. No person can
see girls of humble education, under the influence of music while in the
nervous sleep, without perceiving, that those individuals, if awake, could
not move with the elegance they exhibit under that influence. The
reason of such grace probably is, that it arises from the simple and pure
efforts of nature to balance the body perfectly in all its complicated movements while the power of sight is suspended.

inclination to *general muscular effort*, as well as laughter; in the former there seems to be no inclination to *any* bodily effort, unless excited by *impressions from without*. When the latter are used, there is a remarkable difference again in the power of locomotion and accurate balancing of themselves, when contrasted with the condition of intoxication from wine or spirits, where the limbs become partially paralyzed, whilst the judgment remains pretty clear and acute. The state of muscular quiescence, with acute hearing, and dreamy, glowing imagination, approximates it somewhat to the condition induced by conium.

During the course of last spring some lectures were delivered in this town to prove that the *mesmeric phenomena* might be induced by an "undue continuance or repetition of the same sensible impression" on *any* of the senses. Immediately after the first lecture I instituted experiments according to this plan, but very soon ascertained, that the sleep induced by this mode of operating, *unless through* the *eye*, was nothing more than NATURAL or common sleep, *excepting in patients who had had the impressibility stamped on them, by having been previously mesmerized or hypnotized.* The lecturer concluded his course by stating his opinion, that he knew no sleep but natural or common sleep; and by representing that he considered the effects produced by the different modes to be the same.[1]   I

---

[1] This being his belief, there could be no novelty in his views. The following was the language of Cullen, long before he was born, "If the mind is attached to a single sensation, it is brought very nearly to the state of the total absence of impressions; or, in other words, to the state most closely bordering upon sleep; remove those stimuli which keep it employed, and sleep ensues at any time."

M'Nish also writes, "Attention to a single sensation has the same effect (of inducing slumber). This has been exemplified in the case of all

believe most, if not all the patients this gentleman
exhibited at his lectures, had been previously mes-

kinds of monotony, where there is a want of variety to stimulate the
ideas, and keep them on the alert."

And again M'Nish writes, "I have often coaxed myself to sleep by
internally repeating half a dozen times any well-known rhyme. Whilst
doing so the ideas must be strictly directed to this particular theme, and
prevented from wandering." He then adds, that the great secret is to
compel the mind to depart from its favourite train of thought, into which
it has a tendency to run, "and address itself solely to the *verbal* repetition
of what is substituted in its place ;" and farther adds, "the more the mind
is brought to turn upon a *single impression*, the more closely it is made
to approach to the state of sleep, which is the total absence of all impressions." Willich also, some forty years ago, wrote thus, "Sleep is promoted by tranquillity of mind, . . . by *gently and uniformly affecting
one of the senses ;* for instance, by music or reading ; and lastly, a gentle
external motion of the whole body, as by rocking or sailing." Counting
and repeating a few words have been also long and generally known and
resorted to for the purpose of procuring sleep.

Let any one read attentively the following extract from the *Medical
Gazette* of February 24, 1838, on the power of weak monotonous impressions on the senses having the power of inducing sleep, and many
phenomena usually attributed to mesmerism, and say what merit could be
due to a person acquainted with the article referred to, for *recording a
note* to *the same effects some six or eight months thereafter*, and that without
having instituted a single experiment to prove the correctness of the
hypothesis ? "For the other slight symptoms" (others enumerated having
been attributed to imagination or emotion of mind) "of vapours, drowsiness, and at last natural sleep, no other cause need be sought than the
tediousness and ennui of passing the hands for more or less than an hour
over the most sensitive parts of the body. This is only an instance of the
well-known effect of weak monotonous impressions on the senses inducing
sleep ; analogous examples are found in the soothing influence of a body
seen slowly vibrating, or of a distant calm scene, or the motions of the
waves, or of quivering leaves ; or in impressions on the sense of hearing
by the sound of a waterfall, the rippling of billows, the humming of insects,
the low howling of the winds, the voice of a dull reader ; or on the
nerves of common sensation by *gentle friction of the temple or eye-brow,
or any sensitive part of the body ;* the rocking of a cradle ; any slow and
regular motion of the limbs or trunk ; all these instances show that the
effect of monotonous impressions on the senses is to produce, in most
persons, tranquillity, or drowsiness, and ultimately sleep."

Where, then, is the great merit of any one having recorded a note six
or eight months after this was published, that these phenomena were
induced by "the undue continuance and repetition of the same sensible
impression ? "

merized or hypnotized, which, if I am correct in this supposition, from the circumstances already referred to (see page 86, and note hereto appended), would completely nullify the importance of his *apparent* results. However, I have never heard of his having *operated successfully, and exhibited the phenomena on numbers of patients taken indiscriminately from a mixed audience, who had never been operated on before;* or produced curative results such as I have so repeatedly done. I therefore consider it a fair inference, that until the same phenomena are produced by his method in cases of persons which have *never* been hypnotized or mesmerized, nothing is proved beyond the fact *which I have so often urged,* namely, the power of imagination, sympathy, and habit, in producing the expected effects ON THOSE PREVIOUSLY IMPRESSED.[1]

From overlooking another important fact which I have repeatedly explained, that all the phenomena are consecutive, that is, first increased sensibility, mobility, and docility, and afterwards a subsidence into insensibility and cataleptiform rigidity, this gentleman, by mistaking and exhibiting the *primary* phenomena for

---

[1] A very decided proof of this was exhibited at one of my lectures, where, as may be seen from the report of it, twenty-two who had been operated on before, laid hold of different parts of each other's persons or dresses, and by concentrating their attention to that act, and anticipating the effect, they all became hypnotized in about a minute. After another lecture, in the ante-room, sixteen who had been hypnotized formerly, stood up in the same manner, and also *one* who *had never been hypnotized.* In about a minute all were affected *excepting the latter.* I then operated on him alone in my usual way, and in two or three minutes he was very decidedly affected. Suffice it to say, I have varied my experiments in every possible form, and clearly proved the power of imagination *over those previously impressed,* as the patients have become hypnotized or not by the same appliance, accordingly to the result which they previously expected. This readily accounts for the result of Mr Wakley's experiments with the Okeys.

the *secondary*, seems to have managed to deceive both himself and some others who are satisfied to look at such matters loosely. *This, however, is confounding things which are in themselves essentially different.* I beg especial attention to the note below.[1]

[1] In illustration of this, I may here state the following remarkable facts, which have been frequently repeated before many most competent witnesses, and of which, therefore, I consider there can be no doubt.

The first symptoms after the induction of the hypnotic state, and extending the limbs, are those of extreme excitement of all the organs of sense, sight excepted. I have ascertained by accurate measurement, that the hearing is about twelve times more acute than in the natural condition. Thus a patient who could not hear the tick of a watch beyond 3 feet when awake, could do so when hypnotized at the distance of 35 feet, and walk to it in a direct line, without difficulty or hesitation. Smell is in like manner wonderfully exalted ; one patient has been able to trace a rose through the air when held 46 feet from her. May this not account for the fact of Dr Elliotson's patient Okey, discovering the peculiar odour of patients *in articulo mortis?* when she said on passing them, " there is Jack." The tactual sensibility is so great, that the slightest touch is felt, and will call into action corresponding muscles, which will also be found to exert a most inordinate power. The sense of heat, cold, and resistance, are also exalted to that degree, as to enable the patient to feel any thing *without actual contact*, in some cases at a considerable distance (18 or 20 inches), if the temperature is very different from that of the body ; and some will feel a breath of air from the lips, or the blast of a pair of bellows, at the distance of 50, or even 90 feet, and bend from it, and, by making a back current, as by waving the hand or a fan, will move in the opposite direction. The patient has a tendency to *approach to, or recede from impressions, according as they are agreeable or disagreeable, either in quality or intensity.* Thus, they will approach to soft sounds, but they will recede from loud sounds, however harmonious. A discord, such as two semi-tones sounded at same time, *however soft*, will cause a sensitive patient to shudder and recede when hypnotized, although ignorant of music, and not at all disagreeably affected by such discord when awake. By allowing a little time to elapse, and the patient to be in a state of quietude, he will lapse into the opposite extreme, of rigidity and torpor of *all* the senses, so that he will not hear the loudest noise, nor smell the most fragrant or pungent odour ; nor feel what is either hot or cold, although not only approximated to, but brought into actual contact with, the skin. He may now be pricked, or pinched, or maimed, without evincing the slightest symptom of pain or sensibility, and the limbs will remain rigidly fixed. At this stage a puff of wind directed against any organ *instantaneously* rouses it to inordinate sensibility, and the rigid muscles to a state of mobility. Thus, the patient may be unconscious of the loudest

Of all the circumstances connected with the artificial
sleep which I induce, nothing so strongly marks the

noise, but by simply causing a current of air to come against the ear, a
very moderate noise will *instantly* be heard so *intensely* as to make the
patient start and shiver violently, although the whole body had immediately before been rigidly cataleptiform.   A rose, valerian, or asafœtida, or
strongest *liquor ammoniæ*, may have been held close under the nostrils
without being perceived, but a puff of wind directed against the nostrils
will instantly rouse the sense so much, that supposing the rose had been
carried 46 feet distant, the patient has instantly set off in pursuit of it ;
and even whilst the eyes were bandaged, reached it as certainly as a dog
traces out game ; but, as respects valerian or asafœtida, will rush *from*
the unpleasant smell, with the greatest haste.   The same with the sense
of touch.

The remarkable fact that the whole senses may have been in the state
of profound torpor, and the body in a state of rigidity, and yet by very
gentle pressure over the eye-balls, the patient shall be instantly roused to
the waking condition, as regards all the senses, and mobility of the head
and neck, in short to all parts supplied by nerves originating above the
origin of the fifth pair, and those inosculating with them, and will not be
affected by simple mechanical appliance to other organs of sense, is a
striking proof that there exists some remarkable connection between the
state of the eyes, and condition of the brain and spinal cord during the
hypnotic state.

This is also a remarkably good illustration of the propriety of Mr Mayo's
designation of the origin of the fifth pair of nerves, which he styles " the
dynamic centre of the nervous system."[1]

Another remarkable proof to the same effect is this ; supposing the
same state of torpor of all the senses, and rigidity of the body and limbs
to exist, a puff of air, or gentle pressure against ONE eye will restore sight
to *that eye*, and sense and mobility to *one half of the body*—the same side as
the eye operated on—but will leave the other eye insensible, and the other
half of the body rigid and torpid as before.   Neither hearing nor smell,
however, are restored in this case to either side.   Thus, by one mode of
acting through the eye, we reduce the patient to a state of hemiplegia, by
the other to that of paraplegia, as regards both sense and motion.   In
many cases, when the patient has been hypnotized by looking sideways,
this gives a tendency to the body to turn round in that direction when
asleep.

It seemed puzzling, that by acting on one eye, both sense and motion
could be communicated to the *same* side of the body, seeing the motor
influence is communicated from the *opposite hemisphere of the brain.*   It
has occurred to me that the partial decussation of the optic nerves may
account for this, and that this partial decussation may be for the express

---

[1] " The Nervous System, and its Functions," p. 27.

difference between it and *natural* sleep as the wonderful
power the former evinces in curing many diseases of

purpose of perfecting the union of sensation and motion through the eyes,
"on which we lean as on crutches;" thus enabling us to balance ourselves so much more perfectly than we could otherwise have done.

There is another most remarkable circumstance, that whilst the patient
is in the state of torpor and rigidity, we may pass powerful shocks of the
galvanic battery through the arms, so as to cause violent contortions of
them, without his evincing the slightest symptom of perceiving the shocks,
either by movement of the head or neck, or expression of the countenance.
On partially arousing the head and neck, as by gentle pressure on the eyes,
or passing a current of air against the face, the same shocks *will* be felt, as
evinced by the movements of the head and neck, the contortions of the
face, and the whine, moan, or scream of the patient. All this may happen,
as I have witnessed innumerable times, and the patient be altogether
unconscious of it when roused from the hypnotic condition.

Moreover, whilst the patient is in the condition to be unconscious of
the shock passed through the arms whilst a rod is placed in each hand, if
one of the rods is applied to any part of the head, or neck, or face, in
short, to any part which is set at liberty by acting on *both eyes*, as formerly
referred to, he will instantly manifest symptoms of feeling a shock, though
it be much less powerful than that which had failed to produce any sensation or consciousness when passed through both arms. This might readily
be accounted for on the principle of the circuit being shortened, and also
by one of the rods being nearer the centre of the sensorium ; but that it
depends on something else is apparent from the following fact : Without
moving the rod placed on the neck, head, or face, carry the other rod
from the hand, to any other part of the head, neck, or face, and all
evidence of feeling will disappear, *unless the power of the galvanic current
is increased.*

Analogous to this is another most puzzling phenomenon : The brain
being in a state of torpor, the limbs rigid, and the skin insensible to
pricking, pinching, heat or cold, by gently pressing the point of one or
two fingers against the back of the hand, or any other part of the extremity, the rigidity will very speedily give place to mobility, and quivering of the arm, hand, and fingers, and which is greatly increased by pressing another finger against the neck, head, or face. Indeed, in the latter
case, the commotion of the whole body is as violent in some patients as
from shocks of the galvanic battery. By placing BOTH fingers on any
part of the head, face, or neck, the commotion almost, or entirely ceases.
By pinching the skin of the hand or arm with one finger and thumb, and
the skin of the neck or face with the other, no effect is produced. Pressure, made with insulating rods, glass, or sealing wax, is followed by the
same phenomena as when done by the points of the fingers. The flat hand
applied has very little effect. The pressure being made against both
hands, the arms are contorted, and if the head is partially dehypnotized

long standing, and which had resisted *natural* sleep, and every known agency, for years,—*e.g.* patients who have been born deaf and dumb, of various ages, up to 32 years, had continued without the power of hearing sound until the time they were operated on by me, and yet they were enabled to do so by being kept in the hypnotic state for eight, ten, or twelve minutes, and

the patient will complain of pins running into the fingers, especially if one point of contact is the hand, and the other the face or head. These phenomena do not occur whilst the skin remains sensible to pricking or pinching.

Moreover, during the state of cataleptiform rigidity, the circulation becomes greatly accelerated, in many cases it has more than double the natural velocity ; and may be brought down to the natural standard, in most cases in less than a minute, by reducing the cataleptiform condition. It is also found, that it may be kept at any intermediate condition between these two extremes, according to the manipulations used ; and that the blood is circulated with less *force* (the pulse being always contracted) in the *rigid limbs*, and sent in correspondingly greater quantity and force into those parts which are not directly subjected to the pressure of rigid muscles. It is also important to note, that by acting on both eyes in the manner required to induce the state of paraplegia, as already explained, the force and frequency of the heart's action may be as speedily and perceptibly diminished, as the action of a steam engine by turning off the steam. By again fixing the eyes, its former force and velocity will be almost as speedily restored, as can be satisfactorily proved to any one who keeps his ear applied to the chest during these experiments. The amount of change in the pulse, by acting on the two eyes, and thus liberating the organs of special sense, and the head and neck, is about 60 per cent. of the actual rise of the pulse when at the maximum above the ordinary velocity of the circulation. We might therefore, I think, *a priori*, infer, that in this new condition of the nervous system we have acquired an important power to act with.

N.B.—It is to be observed, that owing to the extreme acuteness of hearing during the first stage of hypnotism, it is extremely apt to mislead the operator, or those who do not understand this fact, during operations on the acuteness of the other senses, such as smell, currents of air, and heat and cold. To avoid such mistakes, therefore, it is best to allow the hearing to disappear, by which time all the other senses will have gone to rest, with the exception of the susceptibility to be affected by a current of air. I allow all the senses to become dormant, and then rouse only the one I wish to exhibit in the state of exalted function, when operating carefully.

have had their hearing still farther improved by a repetition of similar operations. Now, supposing these
patients to have spent six hours out of twenty-four in
sleep, many of them had had four, five, six, or eight
years of *continuous* sleep, but still awoke as they lay
down, incapable of hearing sound, and yet they had
some degree of it communicated to them by a few
minutes of Hypnotism. Can any stronger proof be
wanted, or adduced, than this, that it is very different
from *common* sleep? A lady, 54 years of age, had been
suffering for 16 years from incipient amaurosis. According to the same ratio, she must have had four years
of sleep, but instead of improving she was every month
getting worse, and when she called on me, could with
difficulty read two words of the largest heading of a
newspaper. After *eight minutes* hypnotic sleep, however, she could read the other words, and in three
minutes more, the whole of the smaller heading, soon
after a smaller sized type, and the same afternoon, with
the aid of her glasses, read the 118th Psalm, 29 verses,
in the small diamond Polyglot Bible, which for years
had been a sealed book to her. There has also been
a most remarkable improvement in this lady's general
health since she was hypnotized. Is there any individual who can fail to see, in this case, something different from common sleep? Another lady, 44 years of
age, had required glasses 22 years, to enable her to see
to sew, read, or write. She had thus five years and
a-half of sleep, but the sight was still getting worse, so
that, before being hypnotized, she could not distinguish
the capitals in the advertising columns of a newspaper.
After being hypnotized, however, she could, in a few
minutes, see to read the large and second heading of

the newspaper, and next day, to make herself a blond cap, threading her needle WITHOUT the aid of glasses. This lady's daughter, who had been compelled to use glasses for two years, was enabled to dispense with them, after being *once* hypnotized. It is also important to note, that all these three, as well as many others, were agreeably surprised by improvement of *memory* after being hypnotized. The memory of one was so bad that she was often forced to go up stairs several times before she could remember what she went for, and could scarcely carry on a conversation; but all this remnant of a slight paralytic affection is gone, by the same operations which roused the optic nerves, and re-stored the sight. Now, with such cases as these, who can doubt that there is a real difference in the state of the brain and nervous system generally, during the hyp-notic sleep, from that which occurs in common sleep? The same might be urged from various other diseases cured or relieved by this process, but I shall only briefly refer to a few.

In the second part of this treatise, where the cases are recorded, will be found many examples of the curative power of hypnotism, equally remarkable with those to which I have just referred: such as Tic Doloureux; Nervous headache; Spinal irritation; Neuralgia of the heart; Palpitation and intermittent action of the heart; Epilepsy; Rheumatism; Paralysis; Distortions and tonic spasm, etc.

I shall here give a few particulars of a case which shows in a most remarkable degree the difference of this and common sleep, or that induced by opium and the whole range of medicines of that class. Miss Collins, of Newark, Nottinghamshire, had a spasmodic seizure during the night, by which her head was bound firmly to her left shoulder. The most energetic and well directed means, under a most talented physician, and aided by the opinion of Sir Benjamin Brodie, had been tried, as far as known remedies could be carried (amongst other means, narcotics, in as large doses as were compatible with the safety of the patient); and although she was carefully watched by night and by day, there had never been the slightest relaxation of the spasm, which had continued nearly six months. When I first examined her, no force I was capable of exerting could succeed in separating the head and shoulder in the slightest degree. Experience led me to hope, however, that I might be able to do so after she was hypnotized. Having requested all present, excepting the patient, her father, and her physician, to retire, I hypnotized her, and in three minutes from commencing the operation, with the most perfect ease to myself, and without the slightest pain to the patient, her head was inclined in the opposite direction, and in two minutes more she was roused, and was quite straight. I visited this patient only three times, after which she returned home. Shortly afterwards, she had a nervous twitching of the head, and on one occasion it was again drawn down to her shoulder. Dr Chawner, however, hypnotized her as he had seen me do, and put it right immediately; and she is now (about twelve months after she was hypnotized) in perfect health, ("her head quite straight, and she has perfect control over the muscles of the neck." (See cases.)

Miss E. Atkinson had been unable to speak above a whisper for four years and a half, notwithstanding every known remedy had been perseveringly adopted, under able practitioners. After the ninth hypnotic operation she could speak aloud without effort, and has continued quite well ever since—now about nine months. (See case at length, Part II.)

The extraordinary effects of a few minutes hypnotism, manifested in such cases (so very different from what we realise by the application of ordinary means) may appear startling to those unacquainted with the remarkable powers of this process. I have been recommended, on this account, to conceal the fact of the rapidity and extent of the changes induced, as many may consider the thing *impossible*, and thus be led to reject the *less* startling, although *not more* true, reports of its beneficial action in other cases. In recording the cases, however, I have considered it my duty to record *facts as I found them*, and to make no compromise for the sake of accommodating them to the preconceived notions or prejudices of others.

It may be proper to add, however, that I have afforded opportunities to many eminent professional and scientific gentlemen to see the patients, and investigate for themselves the real state of these respective cases; and to them I can confidently appeal as to the accuracy and fidelity of the reports of most of the cases recorded in this treatise.

After such evidence as this, no one can reasonably doubt that there is a remarkable difference between hypnotism and natural sleep, and that it is a valuable addition to our therapeutic means.

How these extraordinary effects are produced, it may be impossible absolutely to decide. One thing, however, I am certain of, that, in this condition, besides the peculiar impression directly made on the nervous centres, by which the mind is for the time "thrown out of gear," and which enables us, in a remarkable manner, to localize or concentrate the nervous energy, or sensorial power, to any particular point or function, instead of the more equal distribution which exists in the ordinary condition, we have also an extraordinary power of acting on the capillaries, and of increasing and diminishing the force and frequency of the circulation, locally and generally.[1] This can be done in a most remarkable degree, both as regards the extent and rapidity of these changes.[2] And, moreover, changes from absolute insensibility to the most exalted sensi-

[1] By this I mean that any one examining the pulse by the radial artery, whilst the patient has his arms in the cataleptiform condition, and held at right angles with his body, (and when, of course, the circulation can only be influenced by the state of rigidity or flaccidity of the muscles), it will be found feeble or contracted, but the moment the rigidity of the muscles is reduced, by blowing on or fanning them, the pulse will become much more developed.   This, of course, which may be done without the patient being conscious of the experiment, is totally different from what may be displayed as a trick, by a person voluntarily compressing the axillary and brachial arteries, by drawing his arm firmly against his side.   The former is independent of volition, the latter is entirely voluntary, and a mere trick.

[2] The first time I ever had an opportunity of examining a patient minutely, or of feeling the pulse of one, under the mesmeric influence, was on the 19th November, 1841.   I was much struck with the state of the pulse at the wrist—so small and rapid as, combined with the state of tremor, or slight subsultus in the arm, rendered it impossible to count it accurately at the wrist.   This circumstance induced me to reckon the velocity of the pulse by the carotid artery, as will be found recorded in the *Manchester Guardian* of the 24th of that month.   I adduced this as the cause of the discrepancy between the numeration of the pulse by others and myself, that I had counted it *by the carotid artery*, and considered it impossible for any one to reckon it correctly by the radial artery in such a case.   The injected state of the conjunctival membrane of the eye, and the whole capillary system in the neck, head, and face, was so apparent, as Dr Radford very correctly stated, that no one near the patient could fail to observe it : this, together with the cold hands and contracted pulse at the wrist, led me to infer, that the rigid state of the cataleptiform muscles, opposed the free transmission of the blood through the extremities, and

bility, may be effected at a certain stage, almost with the rapidity of thought, as exemplified at page 136.   On the whole, I consider it is of great importance to have acquired a knowledge of how these effects can be produced and generally applied, and turned to advantage in the cure of disease, although we should never ascertain the real proximate cause, or principle through which we produce our effects.   Who can tell how, or why, quinine and arsenic cure intermittent fever?   They are, nevertheless, well known to do so, and are prescribed accordingly.

Whilst I feel assured from personal experience, and the testimony of professional friends, on whose judgment and candour I can implicitly rely, that in this we have acquired an important curative agency for *a certain class* of diseases, I desire it to be distinctly understood, as already stated, that I by no means wish to hold it up as a universal remedy.   I believe it is capable of doing great work, if judiciously applied.   Diseases evince totally different pathological conditions, and the treatment ought to be varied accordingly.   We have, therefore, no right to expect to find a universal remedy in *this*, or *any other*, method of treatment.

would thus cause increased action of the heart and determination to the brain and spinal chord, as resulted from the ingenious experiments of my late friend Dr Kellie, for speedily terminating the cold stage of ague, by putting a tourniquet round one of the extremities.

K

CHAPTER V

WHEN I had ascertained that Hypnotism was important as a curative power, and that the prejudices existing against it in the public mind, as to its having an immoral tendency, were erroneous ; and the idea, that it was calculated to sap the foundation of the Christian creed, by suggesting that the Gospel miracles might have been wrought by this agency, was quite unfounded and absurd, I felt it to be a duty I owed to the cause of humanity, and my profession, to use my best endeavours to remove those fallacies, so that the profession generally might be at liberty to prosecute the inquiry, and apply it practically, without hazarding their personal and professional interest, by prosecuting it in opposition to popular prejudice. It appeared to me there was no mode so likely to insure this happy consummation as delivering lectures on the subject to mixed audiences. The public could thus have demonstrative proof of its practical utility ; and, when it was proved to proceed from a law of the animal economy, and that the patient could only be affected *in accordance with his own free will and consent*, and not, as the animal magnetizers contend, through the irresistible power of volitions and passes of the mesmerizers, which might be done in secret and at a distance, the ground of charge as to my agency having an immoral tendency, must at once fall to the ground. I have reason to believe my

labours have not been altogether unsuccessful, in removing the popular prejudices; and I hope that the more
liberal of my professional brethren, now that they know
my true motives of action, in giving lectures to mixed
audiences, instead of confining them to the profession
only, and *especially as I made no secret of my modes of
operating*, will be inclined to approve rather than blame
me, for the course I have taken in this respect.

From some peculiar views, I was led to make experiments, by which I hoped to obtain natural or refreshing
sleep, and the results were quite satisfactory. I have
thus succeeded in making a patient, who, *when operated
upon in the usual way*, was highly susceptible, and disposed to become strongly cataleptic, with rapid pulse
and oppressed breathing, remain in a sound sleep for
upwards of three hours, with all the muscles flaccid, and
the pulse and respiration slower than natural, when
operated on in this manner. All this difference arises
from the simple circumstance of the *position in which the
eyes are placed during the operation*, namely, closing the
eyelids, and bringing the eyes loosely upwards, as if
looking at an object at a great distance, the eye-balls
being turned up only *gently*, so as to cause *dilatation* of
the *pupil*, as already explained; and the limbs placed so
as to relax the muscles as much as possible, and thus
prevent acceleration of the pulse.

I was led to the adoption of this method from the
following train of reasoning. If, as I inferred was the
case, the spasmodic tendency was reflected to the muscular system generally, from the semi-paralyzed state
of the branches of the third pair of nerves (which
supply the levatores palpebrarum and irides), during
the continued fixed stare and straining of the eyes, I

thought, were I to insure all the *other* concomitant
requirements for procuring hypnotism, *minus* the *strain
on the levators and irides*, I ought to procure refreshing
sleep, without rigidity of muscle or quickened circulation.   By closing the eyelids, the first could be obtained, and by turning the eyes up loosely, which
dilates the pupils, the other would also be attained ; I
therefore tried the experiment, which, as already noted,
proved most successful.

I think the plan I have just pointed out is quite as
simple, and I feel assured it will prove as efficacious
in procuring " sleep at will " as that of Gardener, lately
published by Dr Binns.   I may add, that I publicly
stated my plan at my lectures in London, on the 1st
and 2nd March, 1842, which was at least five or six
months prior to the publication of Dr Binn's work.   I
had also done the same at my lectures in Liverpool,
about six weeks before that last period.

Mr Barrallier, an intelligent surgeon, of Milford, who
investigated the subject of Hypnotism with much zeal
and success, and published some interesting experiments on the subject in the *Medical Times*, also referred
to the case of a gentleman in that town, whom he had
heard of as having been in the habit of procuring
sleep immediately by keeping his eyes fixed for a few
minutes in one direction.   Until he adopted this method
he scarcely slept at all.   For various modes of procuring sleep see pages 133, 134 of this treatise.

In reference to my original theory, Dr Binns, at page
372, calls in question the justice of my allegation, that
during Hypnotism, natural or artificial, there should
be any imperfect arterialization of the blood, notwithstanding the suppressed or modified respiration and

circulation. He has adduced no arguments, however, to convince me to the contrary; and I again repeat my conviction, that such condition of the blood *does* exist, and is a cause of ordinary sleep; and that the still more intense state of torpor, in a certain stage of Neuro-Hypnotism, results from a still less perfectly purified blood ; and, on the other hand, that the dreamy and exalted states arise from different degrees of stimulating properties of the blood, from being more highly arterialized at various stages, together with the velocity of circulation, and pressure or tension on the brain during the cataleptiform state.

CHAPTER VI

I HAVE no doubt that some of the views already advanced, and the facts on which they are grounded, have appeared startling to many of my readers, and I feel assured the subject to be discussed in the following chapter must be still more so; namely, that during hypnotism, we acquire the power, through the nerves of common sensation, of rousing any sentiment, feeling, passion, or emotion, and any mental manifestation, according to our mode of manipulating the patient. This is what has been designated phreno-magnetism by the discoverers of these curious phenomena, but which, in accordance with my nomenclature, I shall designate phreno-hypnotism. It appears with this, as with many other discoveries, that similar investigations were going forward at the same time in England and America, while the discoverers were without the knowledge of each other's views or proceedings, and that the results of their experiments led all parties to form analogous conclusions.

It must be evident to every one who reflects deeply and dispassionately on the subject, that if we really can thus acquire such power as to rouse into great activity any faculty or propensity, whilst we diminish the activity of antagonist faculties, we must thereby acquire an important power for meliorating the moral, intellectual, and physical condition of man. I shall have no difficulty in adducing sufficient proof, that the human mind *can* be so developed and acted on through the bodily organs; but, before entering into a detail of the modes of doing so, I shall endeavour to remove a prejudice against the discussion of this subject, which has arisen from the unhappy circumstance that some of those who promulgated this doctrine have professed a belief in materialism. Such an avowal was indeed calculated to excite, not the prejudices, but the sound principles, of Christian society in general against the reception or dispassionate consideration of the facts on which it rested. For my own part, I can see nothing in the subject to warrant such conclusions as the materialists have avowed; and truth is not to be rejected, because misguided men attempt to build upon it a hollow and unseemly superstructure.

The following are my views of the relation which subsists betwixt mind and matter :—I look upon the brain simply as the *organ* of the mind, and the bodily organs as the instruments for upholding the integrity of the bodily frame, and for acquiring and extending its communion with external nature in our present state of existence. That the mind acts *on* matter, and is acted on *by* matter, according to the quality and quantity, and relative disposition of cerebral development. This, however, does not imply, that mind is *a mere attribute of matter*.[1] My thinking, and willing,

---

[1] "A few sounds acting on the tympanum of the ear, or a few black and small figures scribbled on a piece of white paper (see Mr Rennell's pamphlet) have been known to knock a man down as effectually as a sledge hammer, and to deprive him not only of vision, but even of life. Here, then, we have instances of mind acting upon matter, and I by no means affirm that matter does not also act upon mind ; for to those who

and acting, so as to influence the mental and bodily condition of another, surely does not destroy our separate individuality? As well might we say, that the refined compositions of a Mozart or Beethoven, which were conveyed to the ears of their delighted auditors through different instruments, were created by the thought and will of the instruments.

It appears to me quite clear, that the musician might conceive, and compose, and record every idea, whilst others could have no conception of their nature or merits, unless communicated *through an appropriate instrument or instruments*. The musician and instrument, therefore, are distinct in their nature, as the soul and the bodily organs are essentially distinct from each other.

I shall endeavour to illustrate my views by the

advocate the intimate connection between body and mind, these reciprocities of action are easily reconcilable ; but this will be an insuperable difficulty to those who affirm the identity of mind and body." Again, "This intimate union between body and mind is, in fact, analogous to all that we see, and feel, and comprehend. Thus, we observe that the material stimuli of alcohol, or of opium, act upon the mind through the body, and that the moral stimuli of love, or of anger, act upon the body through the mind : these are reciprocities of action that establish the principle of connection between the two, but are fatal to that of an identity."—"Does not every passion of the mind act directly, primarily, and, as it were, *per se* upon the body, with greater or with lesser influence in proportion to their force ? Does not the activity belong on this occasion to the mind, and the mere passiveness to the body? Does not the quickened circulation *follow* the anger, the start the surprise, and the swoon the sorrow? Do not these instances, and a thousand others, clearly convince us that priority of action *here* belongs to the mind, and not to the body? and those who deny this, are reduced to the ridiculous absurdity of attempting to prove that a man is frightened because he runs away, not that he runs away because he is frightened, and that the motion produces the terror, not the terror the motion."—Colton's *Lacon*. The same author also urges the argument effectively by an appeal to the fact of mania being so frequently produced by *moral* causes, and the success which has attended the treatment of the insane by strict attention to *moral* management.

following simile. Suppose the instrument is good, and well fitted for expressing musical composition, it is evident, that it will better convey the beauties of the composition, than if represented by a *bad* or *indifferent instrument;* and will also afford more delight, and satisfaction, and encouragement to the farther exertions of the composer, than if performed on a bad instrument. Just so the mind furnished with a well developed brain. Supposing the musical instrument is very perfect in *some* parts, but very *im*perfect in others, it is evident, that the musician can afford more pleasure to others, as well as more satisfaction to himself, by playing on the more perfect parts. Then, supposing the parts played on capable of becoming improved, *by being so exercised* (which is the case with several instruments, as the violin), it is clear, that there will be greater and greater inducement for the musician to confine himself to the better parts of the instrument, and thus, by concentrating his whole energies to these points, he will more and more enamour himself, as well as his auditors, by the perfection of his performances.

This is exactly what I conceive takes place in reference to the brain, supposing different parts to be appropriated as the instruments for the manifestation of different mental functions. Every part of the human frame is continually undergoing the process of waste and repair—that is to say, the molecular particles of the various organs are continually changing, and *moderate* exercise tends to *increased development and power*, whilst *inaction* has the opposite tendency. This no one will deny. The analogy, therefore, is complete. The soul or mind, by being exercised judiciously in a particular direction, strengthens some peculiar organ,

and acquires precision from habit, which gives a tendency to perseverance in the same course of action ;
and, by refraining from certain practices, the corresponding organs become feeble, and thus exercise a
less powerful influence on the mind. Thus we can
account for the power of habit, both physical and
mental, each tending to strengthen the other by correct
training ; and it is on this principle that we can hope
to meliorate the condition of the vicious members of
society, by separating them from bad companions
and practices, and encouraging them in the exercise
of virtuous habits.

Moreover, the mind of the musician may conceive
and excite into activity the corresponding organs of
the brain ; these may react on his corporeal organs,
and excite into activity the silent lyre ; all these links
of intercommunication may be perfected, without conveying any corresponding feeling or emotion to the
minds of others, unless they are provided with appropriate recipient organs (musical ears) for conveying
to their brains certain vibrations, and thus inducing
in corresponding parts of their brains such a condition
as may awaken in their minds certain associations of
ideas, and manifest the peculiar emotions which arise
from them. It is not enough that we have *part* of
this concatenation complete ; the whole must be
complete, or the results cannot be perfect.[1]

[1] Some time after 1 had written the above, I had the satisfaction to
meet with a somewhat analogous illustration from the pen of the late
celebrated Dr John Armstrong, which I now quote from his work on
Fever, p. 478 :—

" It will have been perceived, that I consider insanity as the effect of
some disorder in the circulation, whether produced by agencies of a corporeal or mental nature. It might be shown by familiar facts, that the
brain is the principal organ through which the operations of the mind are

The same arguments might be enforced in respect to the painter, and sculptor, and orator, but it appears

performed ; and it does not, as many have supposed, necessarily involve the doctrine of materialism to affirm, that certain disorders of that organ are capable of disturbing those operations. If the most skilful musician in the world were placed before an unstrung or broken instrument, he could not produce the harmony which he was accustomed to do when that instrument was perfect, nay, on the contrary, the sounds would be discordant ; and yet it would be manifestly most illogical to conclude, from such an effect, that the powers of the musician were impaired, since they merely appeared to be so from the imperfection of the instrument. Now, what the instrument is to the musician, the brain may be to the mind, for aught we know to the contrary ; and to pursue the figure, as the musician has an existence distinct from that of the instrument, so the mind may have an existence distinct from that of the brain ; for in truth we have no proof whatever of mind being a property dependent upon any arrangement of matter. We perceive, indeed, the properties of matter wonderfully modified in the various things of the universe, which strike our senses with the force of their sublimity or beauty ; but in all these we recognize certain radical and common properties, that bear no conceivable relation to those mysterious capacities of thought and of feeling, referable to that something which, to designate and distinguish from matter, we term mind. In this way, I conceive, the common sense of mankind has made the distinction which; every where obtains between mind and matter, for it is natural to conclude, that the essence of mind may be distinct from the essence of matter, as the operations of the one are so distinct from the properties of the other. But when we say that mind is immaterial, we only mean, that it has not the properties of matter ; for the consciousness which informs us of the operations, does not reveal the abstract nature of mind, neither do the properties reveal the essence of matter. When any one, therefore, asserts the materiality of mind, he presupposes, that the phenomena of matter clearly show the real cause of mind, which, as they do not, he unphilosophically places his argument on an assumption. And his ground of reasoning is equally gratuitous, when he contends that mind is an attribute of matter, because it is never known to operate but in conjunction with matter, for though this connection is continually displayed, yet we have no direct proof of its being necessary."

In like manner, Mr Herbert Mayo, in the introduction to his late work on the Nervous System and its Functions, writes thus :—" Life is a force so contrived and used, as to qualify the materials of the inert world for a temporary union with consciousness,—a means how mind may enter into such relations with matter, that it may have its being and part in physical nature, and its faculties developed, and its capabilities and tendencies drawn out and proved (for whatever ulterior purpose) in subjection to, and in harmony with, her laws.

" As we imagine the Supreme Mind to be ubiquitous, infinite, conto my mind so evident by what has already been advanced, that I forbear extending my illustrations, conceiving them to be unnecessary. I therefore conclude that the soul and the brain are essentially quite distinct, and stand much in the same relation to each other as the musician and musical instrument.

Another powerful argument of the mind being an independent essence, is the fact, that amidst the continued changes which we know are going on in the physical frame, we still recognize personal identity; and the remembrance of occurrences, even of early life, after every particle of the body has been changed several times, is reconcilable with the idea of the original mind merely having exchanged and renovated the substance of its dwelling-place; but how can we suppose that each particle had, in retiring, transferred its quantum of knowledge to the particle of matter which was to supply its place?[1]

Colton's remark seems very just when he says,— " Many causes are now conspiring to increase the trunk of infidelity, but materialism is the main root of them all." I have therefore endeavoured, and I hope, by

trolling, but uncontrolled by matter, so in contrast with these attributes we conceive the finite mind to be bound down to place, and to be dependent on a certain arrangement of matter for its manifestation, each power displayed as the property of a tissue, each agency as the function of an organ.

" These views do not lead to materialism. For one cannot disjoin the physiology of the nervous system from mental philosophy, nor investigate the play of its organs without attending to the mind itself. And if equal consideration is given to the two classes of phenomena, it is impossible (so at least it appears to myself) to avoid the conviction, that they are essentially independent the one of the other, and belong to distinct essences; and that ipseity, the consciousness of personal being, is not a mode of material existence, nor physical impenetrability an attribute of that which feels and thinks."

[1] See Appendix I., Note 9.

what has already been said, with some success, to prove,
that the belief in the brain being the organ of the mind,
leads only to the admission of the necessity of certain
conditions of matter, in order to make the varied conditions of mind manifest to ourselves and other beings
with which we are surrounded during the present state
of our existence. The charge against the doctrine of
phrenology, therefore, as leading to a belief in materialism, is altogether unfounded ; for phrenology merely
professes to appropriate to *separate portions* of the brain
the *execution of special functions* or *manifestations*, which
are generally admitted, without hesitation, to result from
its functions as a single organ. I might therefore at
once dismiss the subject, leaving the doctrines of the
existence of a God, and the immortality of the soul,
to the defence of many able writers on that department
of mental philosophy. However, as it appears to me
that an argument of considerable strength, in support
of both these doctrines, may be drawn from the doctrines
of phrenology, or the allocation of special functions to
particular portions of the brain, I think it may not be
out of place for me very briefly to advert to these topics.

The concurrent notions and practices of all nations,
savage as well as civilized, clearly indicate their inward
belief in a superintendent power who rules the destinies
of man and of nations, as verified by their varied forms
of worship. Phrenology, as illustrated by Hypnotism,
does more—it proves that there is a particular portion
of the brain which the mind may use as an organ destined for the special purpose of adoration ; and, as
nothing has been made in vain, or without a final cause,
we may safely infer that such an organ would never
have been made had it not been intended to be ex-

ercised ; and how could it have been exercised worthily
had there been no suitable object of adoration ? The
very fact, therefore, of the existence of such a special
organ having been ascertained, stamps the folly of the
Atheist ; and, as we have proved that mind is not
necessarily a mere attribute of organized matter, but
a distinct essence, we cannot suppose it to be more
perishable than matter ; and as it is an acknowledged
fact, that matter, so far as we can apprehend, is essentially indestructible, analogy would lead us to infer, that
the mind, the more important part of man, will not be
less imperishable ; and, consequently, the most rational
conclusion to which we can arrive is, that the soul is
immortal.

"There is mind, then, as well as matter, or rather, if
there be a difference of the degrees of evidence, there is
mind, more surely than there is matter ; and if at death
not a single atom of the body perishes, but that which
we term dissolution, decay, putrefaction, is only a change
of the relative positions of those atoms, which in themselves continue to exist with all the qualities which they
before possessed, there is surely no reason, from this
mere change of place of the atoms that formed the
body, to infer, with respect to the independent mind,
any other change than that of its mere relation to those
separate atoms.   The continued subsistence of every
thing corporeal cannot, at least, be regarded as indi-
cative of the annihilation of the other substance, but
must, on the contrary, as far as the mere analogy of the
body is of any weight, be regarded as a presumption in
favour of the continued subsistence of the mind, when
there is nothing around it which has perished, and
nothing even which has perished, in the whole material

universe, since the universe itself was called into being."
—Dr Thomas Brown.

" The mind remembers, conceives, combines, and reasons; it loves, it fears, and hopes in the total absence of any impression from without, that can influence, in the smallest degree, these emotions; and we have the fullest conviction that it would continue to exercise the same functions in undiminished activity, though all material things were at once annihilated."—Abercrombie.

Mr Stewart also says, " Of all the truths we know, the existence of mind is the most certain. Even the system of Berkeley concerning the non-existence of matter, is far more conceivable, than that nothing but matter exists in the universe."

Plato also wrote thus:—" The body being compounded, is dissolved by death; the soul, being simple, passeth into another life, incapable of corruption."

That accomplished physician and metaphysician, Dr Abercrombie, after relating the effects on memory of diseases and disorders of the brain, with, in many instances, serious organic lesion, concludes thus: " One thing, however, is certain, that they give no countenance to the doctrine of materialism, which some have presumptuously deduced from a very partial view of the influence of cerebral disease upon the manifestations of mind. They show us, indeed, in a very striking manner, the mind holding intercourse with the external world through the medium of the brain and nervous system; and, by certain diseases of these organs, they show this intercourse impaired or suspended; but they show nothing more. In particular, they warrant nothing in any degree analogous to those partial deductions which form the basis of materialism.   On the contrary, they show us the brain injured and diseased to an extraordinary extent, without the mental functions being affected in any sensible degree." (This power no doubt arises from each hemisphere having corresponding organs, and consequently when only *one* is diseased, the other may be adequate to the manifestation of the mental phenomena.)   " They show us farther, the manifestations of mind obscured for a time, and yet reviving in all their original vigour almost in the very moment of dissolution.   Finally, they exhibit to us the mind, cut off from all intercourse with the external world, recalling its old impressions, even of things long forgotten, and exercising its powers on those which had long ceased to exist, in a manner totally irreconcilable with any idea we can form of a material function." *On the Intellectual Powers*, pp. 154, 155.

In addition to what I have already advanced in refutation of the doctrine of materialism, I beg to submit what appears to me much more probable than that mental manifestations are the result of mere organism, —namely, that organism is the result of mind, or the principle of life influencing or directing organism in accordance with what may be its especial wants and desires.   We know that every seed of a plant has a principle of life imparted to it by the great first cause of all, by which, when sown in congenial soil, it will exert its powers, and appropriate to itself materials from the soil, to form an organism in accordance with its peculiar wants and nature ; and that, having passed through certain conditions, and formed other kindred seed or germs to propagate its kind under a return of favouring circumstances, the plant dies, and is resolved into its original elements. Man and animals also possess similar faculties for propagating and multiplying their species ; and to me it appears far more probable, that the peculiar organism of each variety results from the vivifying or intelligent principle we call life or mind (and no one denies the existence of the former, although we know nothing of its essence or mode of operation), directing and determining appropriate formation, than that the mere accidental union of particles of matter, in definite quantity and form, should be the *cause* of mental phenomena.[1] It is true we can here only speak of analogies, but the analogy in favour of this proposition seems far more natural and probable than the other. Is it not, for example, *à priori*, as probable, on viewing a well planned factory, fitted up with what is called self-acting machinery, for us to suppose that the whole should have been planned, and the machinery constructed, in accordance with the intelligent designs of a skilful artist, as that the brute matter, of which the whole is constructed, came into its particular forms and arrangements of *its own accord, or by accident*, and thus produced the intelligence of the superintending possessor ? The higher and more perfect the original force, or life, or spirit, originally imparted into each species, the more complex and extensive should we expect to find the corresponding organism, to adapt it for the suitable

---

[1] The original identity of structure of the germ of the most various organic beings, constituted, as it always is, of a cell, with a nucleus, seems to prove, that the cause of the variety of classes, families, genera, and species of animals and plants developed from the germ, resides not in the structure or chemical property of the germ, but in the idea or spirit implanted in it at its creation. (*Müller*, p. 1339.)

performance of its more varied functions ; and it therefore was necessary for man to have that superiority,
even in the form and functions of his hands, in which
he so much surpasses that of all other animals, to fit
him for the execution of the more extended range of
operations, which his superior endowments and cerebral
organism fitted him to devise.

In this view of the subject (and it appears to me,
after consulting the various opinions of our ablest
authorities on life and organization, to be the most
satisfactory conclusion I could arrive at), every plant or
animal, however minute, may have a particular vital or
directing principle originally imparted to it, and still
sustained in its power by the great Creator, without the
necessity of according to each an immortal existence
and responsibility. Nor is there any thing irreconcilable
in the supposition that man, with higher original
powers, and more perfect organism, fitting him to use
these appropriately, and who is the highest link in the
chain, in this world, between inorganic matter and the
Supreme Being, should be constituted a responsible
agent, and exist hereafter, whilst those creatures with
less expansive faculties, both of life and organization,
may be exempted from such ultimate responsibility,
and may *not* be immortal.

This is only analogous to what we see in respect
to a commanding officer and his men, the *former only*
being responsible for imprudent enterprises, the latter
being considered merely as instruments in his
hands.

I shall close these remarks on the immortality of the
soul by a quotation from that excellent work, " Abercrombie on the Intellectual Powers." " This momentous

truth rests on a species of evidence altogether different,
which addresses itself to the moral constitution of man.
It is found in those principles of his nature by which he
feels upon his spirit the awe of a God, and looks forward
to the future with anxiety or with hope,—by which he
knows to distinguish truth from falsehood, and evil
from good, and has forced upon him the conviction
that he is a moral and responsible being. This is the
power of conscience, that monitor within, which raises
its voice in the breast of every man, a witness for his
Creator. He who resigns himself to its guidance, and
he who repels its warnings, are both compelled to
acknowledge its power ; and whether the good man
rejoices in the prospect of immortality, or the victim of
remorse withers beneath an influence unseen by human
eye, and shrinks from the anticipation of a reckoning to
come, each has forced upon him a conviction, such as
argument never gave, that the being which is essentially
himself is distinct from any function of the body, and
will survive in undiminished vigour when the body
shall have fallen into decay.

"There is thus, in the consciousness of every man, a
deep impression of continued existence. The casuist
may reason against it till he bewilder himself in his
own sophistries ; but a voice within gives the lie to his
vain speculations, and pleads with authority for a life
which is to come. The sincere and humble inquirer
cherishes the impression, while he seeks for farther
light on a subject so momentous, and he thus
receives, with absolute conviction, the truth which
beams upon him from the revelation of God, that
the mysterious part of his being, which thinks, and
wills, and reasons, shall indeed survive the wreck

of its mortal tenement, and is destined for immortality." [1]

It must be obvious to all, that every variety of passion
and emotion can be excited in the mind by music; but
how does this arise? Simply by the different effects
produced by the varied degrees of velocity, force, quality,
and combinations of the oscillations of the air acting on
the auditory nerves, these again communicated to the
brain, and this acting on the mind and body, creating
corresponding mental and bodily manifestations. Every
one must have observed the remarkable effects evinced
by these means on the physiognomy, and the more
critically observant must have noticed, that in susceptible individuals there is also a very marked change
in the state of the respiration and general posture of
the body. They must also have experienced, in themselves and others, how prone we are to assume a sympathetic condition, both of mind and body, from those
with whom we associate, or during a temporary interview. These physical changes seem to result from a
mental influence imparted through the eyes and ears,
and then reflected from within, through the respiratory,
facial, and spinal nerves, on the external form and
features. Now, such being the case, is there any great
improbability, that by calling the muscles of expression

---

[1] To those who wish to pursue the subject farther, I beg to refer to
Dr Samuel Clarke on the Being and Attributes of God, pp. 70—75;
Jackson on Matter and Mind, pp. 41—47, 51; Warburton's Divine
Legation, vol. I. book 3rd; Drew's Essay on the Immortality of the
Soul; and Ramsay's Principles, pp. 233—5; also Brougham and
Bakewell, where they will find it ably argued as far as Natural Theology
can avail; but the sacred volume contains a lucidity and sanction beyond
all we can adduce from mere human ingenuity, and I therefore conclude
by referring to it, as "life and immortality are clearly brought to light
through the Gospel."

into action during the hypnotic state, by titillating certain nerves, the impression of the feeling with which such external manifestation is generally associated should be reflected on the brain, and excite in the mind the particular passion or emotion? I think it is highly probable this is the true cause of the phrenological manifestations during the hypnotic condition; and as it is the peculiar feature of this condition, that the whole energies of the soul should be concentrated on the emotion excited, the manifestation, of course, becomes very decided. I presume the different points pressed on, through the stimulus given to various fasciculi of nerves, call into action certain combinations of muscles of expression in the face and general frame, and also influence the organs of respiration, and thus the mind is influenced, *indirectly*, through the organs of common sensation and the sympathetic, as sneezing is excited in some by too strong a light irritating the optic nerves. Two patients who are highly intelligent, and remain partially conscious, and who acknowledge they did all in their power to resist the influence excited by manipulating the head, state, that the first feeling was a drawing of the muscles of the face, and affection of the breathing, which was followed by an irresistible impulse to act as they did, but why they could not tell.

In this view of the subject it would resolve itself into the laws of sympathy,[1] and the question then is, where are the external or superficial points of the sympathies located? Experience must decide this, and in the peculiar condition induced by hypnotism, according to my own experience, this can be more readily and certainly determined than in the normal state. These

[1] See Appendix I., Note 10.

points having been ascertained, we can then determine how and where to act according to our particular object; and it can be of no real importance where the cerebral points or special organs may be posited.

As to the real locations of the sympathetic points, by stimulating which we produce peculiar manifestations, they appear to me not to be quite accurately the same in all heads, but, on the whole, pretty near the centres of the organs as mapped out on heads generally approved by phrenologists, and I have had decided proof that there is some relation subsists betwixt the size and function, as in general there is more energy displayed when there is large development, and the negative when it is defective. Thus, a patient with large combativeness or destructiveness, when excited during hypnotism, will display great violence and disposition to attack others, whereas, where they are defective, they will shrink and express a fear that some one is quarrelling, or angry with them.

If the solution of the cause of these remarkable phenomena now given should not be deemed correct, the only other which occurs to my mind as at all satisfactory, is this, that the different fasciculi of sentient nerves excite *directly* the *corresponding points* of the brain, and these again the physical manifestations. We know by what musical combinations and movements we can excite the different passions ; we know also that this arises from some peculiar impression communicated to the brain through the portio mollis of the seventh pair of nerves ; and whether this is conveyed to it as a single organ only, or as a combination of organs, it is clear that, as the origin of the *seventh* is *more remote* from the brain than the origin of the *fifth*, there must,

consequently, be at least as great difficulty in account-
ing for such results being excited through the different
branches of the *seventh* as through those of the *fifth*
pair.

The animal magnetizers do not now contend for *their*
volitions being necessary. Dr Elliotson distinctly states,
in a published letter, dated 11th September, 1842, that
he had "never produced any effect by mere willing";
and adds, "I have never seen reason to believe (and I
have made innumerable comparative experiments upon
the point), that I have heightened the effect of my
processes by exerting the strongest will, or lessened
them by thinking intentionally of other things, and
endeavouring to bestow no more attention upon what I
was about than was just necessary to carry on the process. So far from willing, I have at first had no idea of
what would be the effect of my processes; in exciting
the *cerebral organs*, the effect ensues as well in my
female patient though the manipulator be a sceptic,
and may therefore be presumed not to wish the proper
result to ensue, and though I stand aside, and do not
know what organ he has in view. I have never excited
them by the mere will; I have excited them with my
fingers just as well when thinking of other matters with
my friends, and momentarily forgetting what I was
about," etc. The Doctor also denies his belief in the
phrenological results arising from *sympathy with the*
state of the operator's brain. I feel convinced that he is
right in these sentiments, and believe that the same
degree of mechanical pressure or stimulus to the integuments of the cranium, from an inanimate substance,
when the patient is in the proper stage of the mesmeric
condition, will produce the same manifestation as the

personal touch of either sceptic or believer in animal magnetism. Thus, touching them with a knobbed glass rod, three feet long, has produced the phenomena with my patients as certainly as personal contact, so that if there is any thing of *vital* magnetism in it, it is subject to different laws from that of *ordinary* magnetism or electricity.

Mere pointing I have myself found sufficient to excite the manifestations in several patients, after previous excitement of the organs, but this arises from feeling, as I know the sensibility of the skin in those cases enables them to feel *without actual contact.*

The following experiment seems to me to prove clearly that the manifestations were entirely attributable to the mechanical pressure operating on an excited state of the nervous system. I placed a cork endways over the organ of veneration, and bound it in that position by a bandage passing under the chin. I now hypnotized the patient, and observed the effect, which was precisely the same, for some time, as when no such appliance was used ; after a minute and a half had elapsed, an altered expression of countenance took place, and a movement of the arms and hands, which latter became clasped as in adoration, and the patient now arose from the seat and knelt down as if engaged in prayer. On moving the cork forwards, active benevolence was manifested, and on being pushed back, veneration again manifested itself. I have repeatedly tried similar experiments with this, and other patients, with the like results, including other organs. It is clear there was no mechanical pressure to direct the movement *downwards*, because there was pressure *upwards also ;* and had there been any preconceived notion in the patient's mind, to excite to such action, it ought to have been manifested *immediately* on *passing into the sleep*. None of the patients had the slightest notion of what was my object in making such experiments, and none of them saw the others operated on. At page 208, it will be observed, pressure by their own fingers produced similar manifestations, even whilst their minds were expecting some other results.

Whilst it is generally agreed that the brain admits of being divided into regions for the animal propensities, moral sentiments, and intellectual faculties, it has not been at all possible to prove satisfactorily the exact position and size of each organ, as noted by the phrenologists. Granting that there is a distinct organ or point in the brain for each faculty, which I think is highly probable, still there must ever be insuperable difficulty in thus accurately determining character, even supposing we knew the exact position and size of each organ, because much must depend upon the state of perfection of structure, and activity of the point or organ, as well as its absolute size. Thus, a person with a large eye may have defective sight, whilst a person with a small eye may see clearly and distinctly, the greater perfection of structure, and activity of the optic nerves, more than compensating for mere deficiency of size. So it is with the brain, a part may be abnormally large, and the faculty dull, from want of power or activity, or perfection of organic structure ; and the reverse may obtain, a small development, with high activity, may render its function predominant. It is from a want of such knowledge as this that phrenology must ever prove imperfect, even granting the localities to be correctly ascertained and established. However,

when we have ascertained the points where, by acting
in any peculiar manner, we can excite into activity
particular sympathetic *physical and mental* associations,
whilst the other faculties are put into a state of quiescence, it appears to me to be a matter of far greater
importance, and a subject still more curious, than any
thing ever brought forward by phrenologists.   It is far
more available for practical purposes too.   Phrenologists
could at best only pretend to tell the *natural tendencies*
of an individual, and direct that he should be educated
in accordance with a specific plan, as has hitherto been
done independently of phrenology, from watching the
natural dispositions and habits of different individuals,
by encouraging and directing their studies in such and
such a direction ; but here, *in addition* to this, we have
the power of giving a decided impulse in any particular
direction.   It ought not to be overlooked, that this does
not deprive us of any of our *former* available modes of
instruction in science and morality, but it promises to
prove a powerful auxiliary for expediting and ensuring
the success of those means.   It therefore follows, that
it becomes the duty of every well-wisher to his species
to investigate this matter, and determine how far it is
generally applicable.   It is still more the duty of the
medical faculty to do so, because, should farther experience determine this question in the affirmative, it
is reasonable to expect it may be turned to the best
account in the cure of disease, by applying our remedies
locally, to the cutaneous points which have been ascertained to be the centres of the morbid concatenation.
Thus, leeching and sedatives, etc., might be applied to
such points when there was excitement of the corresponding functions, and *vice versa*, with the reasonable

hope of success ; and if this method cannot be effective, we can be pretty certain of success through hypnotism, by exciting the morbidly low faculty where there is depression, and the antagonist organ where there has been excitement. In this manner I have no doubt but hypnotism may prove of incalculable advantage in the treatment of many cases of insanity, and nervous affections tending to induce that disease.

I am quite aware some will be ready to start an objection to my views, by stating that the scalp, where many or most of these demonstrations have been manifested, is not highly sensitive, that it is not extensively supplied with sentient nerves, and that they all arise from the fifth pair, and do not pass directly through the skull to the subjacent points of the brain. This, however, does not prove that the terminal branches may not ultimately have a special influence on such points, notwithstanding their circuitous course to arrive there. I beg to remind such individuals that we are by no means sufficiently acquainted with the laws and distribution of the nervous system, to be able to prescribe rules as to *how* it *ought* and *must* act. Who does not know, that until the discoveries of our illustrious countryman, Sir Charles Bell, the same nerve was considered to give both sense and motion? And when he propounded that the true cause of its double functions was because of its having *double* roots, was not this announcement scouted for some time, and then, when proved to be true, were not attempts made to rob Sir Charles Bell of the honour of the discovery?

There seems to be great reason to conclude that the distribution of the nerves of the scalp will ultimately be found far more intricate and beautifully arranged than at present we have any conception of.

I shall now proceed to state my views as to the mode in which different parts of the brain are associated with different parts of the body. I have long quite agreed with those physiologists who consider that the *vis nervosa* is something circulated in tubes, that the primitive nerve-tubes do not anastomose, but only run parallel with others, remaining distinct and isolated throughout their course, and that consequently the "cerebral extremity of each fibre is connected with the peripheral extremity of a single nervous fibre only, and that this peripheral extremity is in relation with only one point of the brain or spinal cord : so that, corresponding to the many millions of primitive fibres which are given off to peripheral parts of the body, there are the same number of peripheral points of the body represented in the brain. The sensation of a single point evidently depends on the impression being conveyed by means of a single fibre to a single point of the sensorium." (Müller.) It is from the same cause that we can regulate simple or associated movements of distinct members.

From all these considerations it appears quite reasonable to suppose, and analogy, as respects distinct organs being appropriated for other special functions, warrants the inference, that different parts of the brain may have special functions to perform, both as regards mind and matter ; and that, when such points are excited into inordinate activity, the manifestations will become correspondingly more conspicuous, and *vice versa*.

We know from experience that the various passions and emotions can be excited through the organ of hearing either by music or oratory, through the eyes by painting and sculpture, and likewise, though less extensively and efficiently, through common sensation, and there seems to me to be nothing, *à priori*, to militate against the probability, that this may be effected to a much greater extent than has yet been done, provided we can only discover the peculiar mode of exciting certain portions of the brain. If the views already advanced, that every point of the body supplied by primitive nervous fibres has a distinct corresponding point in the brain, it is clear, that by titillating each peripheral point, we shall excite its corresponding central point ; and from what shall be found detailed in the experiments recorded, it appears highly probable that the respective parts of the brain corresponding to *every* part of the body, may be excited into activity through certain sympathetic points in the integuments of the head and neck, and if so, we may also excite into activity the whole of those actions, mental and muscular, which are associated with each portion of the cerebrum. In this case Smellie's supposition would be completely realized in man. He expresses himself thus :—" I can conceive a superior being so thoroughly acquainted with the human frame, so perfectly skilled in the connection and mutual dependence which subsist between our intellect and our sensitive organs, as to be able, by titillating in various modes and directions, particular combinations of nerves, or particular branches of any single nerve, to excite in the mind what ideas he may think proper. I can likewise conceive the possibility of suggesting any particular idea, or species of ideas, by affecting the nerves in the same manner as these ideas affect them when excited by any other cause." This confident aspiration seems to be now in a great measure realized, by certain modes of manipulating patients during the hypnotic condition, of which I shall now adduce a few illustrations.

My first attempt to excite the phreno-hypnotic phenomena was in the month of April, 1842, in the lecture-room at Liverpool, but it did not succeed. I then tried the experiment repeatedly in private, putting the patients to sleep by contact as well as in my usual way, but still could not succeed. I was anxious to try it fairly, and therefore applied to Mr Brookes, through the kindness of Dr Birt Davies of Birmingham, for information as to the mode Mr Brookes had practised so successfully, and which was most politely communicated to me by both these gentlemen. I tried this mode with several patients, both in my usual plan and that of the animal magnetizers, but was still unsuccessful. I now abandoned it as a hopeless task, presuming the cases which had proved successful with others must have been *lusus naturæ*, or that the operators had deceived themselves, the patients having been led to answer, and give the manifestations they did, from the nature of the leading questions proposed, and might afterwards remember what passed at previous operations, and answer accordingly ; whilst, like natural somnambulists, they might not remember, when awake, what had passed during their sleep.

Last December I was induced to make another attempt, from reading a report of Mr Spencer T. Hall's two first lectures on the subject, at Sheffield: and it was remarkable, that the very first patient I tried in that way exhibited several of the manifestations. However, I was led to refer the result to a totally different cause from what he and the other animal magnetizers did. I concluded it arose from the different degrees of sensibility of different parts of the integuments, conveying correspondingly varied impressions when similarly impressed, and exciting different ideas in the mind, and thus calling up old associations ; and that when similarly impressed the same ideas might again present themselves to the mind. I considered this far more probable than that the brain was affected by any transmission from the operator to the brain directly through the skull ; and to prove this, tried the effect of pressure over parts which had *no cerebral substance directly subjacent*, and the results confirmed my expectations. Thus, pressure on the apex of the *mastoid process*, and the *ossa nasi*, and the *chin*, were as certainly followed by particular manifestations, as pressure on different parts of the cranium were followed by others. I also very soon ascertained that the same points of the cranium, when thus excited, did not excite the *same ideas or emotions* to the minds of *different* patients, which I considered ought to have been the case, according to the views of the stanch phrenologists. I have since discovered the cause of this, namely, *not having operated at* the *proper stage* of the hypnotic condition.

I shall now adduce a few examples. On one subject, after being in the hypnotic condition for a few minutes, by applying gentle pressure over the *ossa nasi*, immoderate laughter was immediately excited, and ceased as abruptly on removing the contact. The abruptness of these transitions, especially from immoderate laughter to the extreme gravity and vacancy of expression peculiar to the hypnotic state, was quite ludicrous, and almost beyond belief. Supposing she were singing the most grave tune and solemn words, the moment the nose was touched in this manner, by any one, she was irresistibly thrown into this merry mood, but would join in the tune again with the utmost gravity the moment the contact ceased. Rubbing the same part, or pinching up the skin over it, seemed to produce no effect whatever. On applying pressure to this patient's chin there was an immediate catch in the breathing, with sighing and sobbing, which would subside on removing the point of contact. By touching both nose and chin at same time there was the most ludicrous combination of laughing and crying, each struggling for the mastery, as we sometimes see in hysteric attacks. Both would cease immediately on removing the contact. Friction or pinching the skin on the chin had no effect of producing such phenomena. In short, no part of this patient which I tested seemed capable of being excited by friction or pinching the integuments, excepting around the orbits, which produced spectra, although less perfectly so than by simple pressure against the bone. This patient, being pressed over the phrenologists' organ of *time*, always expressed a desire " to write "—a letter—to her mother or her brother ; over their organ of *tune*, " to sing " ; between this and wit, " to be judicious " ; the boundary between wit and causality, " to be clever " ; causality, " to have knowledge " ; in the centre of the forehead, to have " a certain perception of learning " ; below this the phrenologists' eventuality, " to be skilful " ; the points of the head occupied by veneration and benevolence were sometimes indicated by the desire " to be virtuous," or " to be honourable " ; most frequently, when the point touched was over benevolence, the answer was, "to be honourable," and when over the other point, "to be virtuous," when both points were touched at same time, it was, "to be honourable and virtuous," and the same answer was always given when these points were touched *combined with No.* 1, *or amativeness.* When the latter was touched alone, the answer always was "to be commended"; when approximating the mastoid process, or over that process, a remarkable placidity, or expression of delight, came over the countenance, and the desire was for "complacency," which, when hypnotized, she defined, "to be civil," but when awake she seemed at a loss to know what the word meant. On touching "combativeness" the placidity of countenance was speedily exchanged for the opposite expression; but on pressure being made immediately above the ears, the most ferocious aspect of countenance was assumed, the breath being suppressed almost to suffocation, the face becoming flushed, with grinding of the teeth; and when the arms were not rigid, the most vigorous efforts at inflicting violence on all who were within her reach, as several gentlemen can attest to their personal knowledge and sorrow. On pressure being applied to the root of the nose, the idea of seeing different forms, and figures, and colours, seemed to be excited in the mind, more vividly when certain points were thus excited, but it could be excited by pressing the integuments against the *under*, as well as *upper* edge of the orbit, with this difference, that the objects seen, or rather spectra excited, were then generally of a painful and distressing character, whereas they were generally of a bright, and glowing, or cheerful

M

description when excited by acting on the *upper* margin of the orbit. I should observe, that care was taken, in all these experiments, *not to press against the globe of the eye.* Thus far the phenomena were pretty uniform in this patient, the answers having been generally very much the same when impressed *exactly in the same way, on the same points, and under similar circumstances in all other respects.* Thus, the last day I had an opportunity of testing this patient, I went over the different points four times with scarcely the slightest variation in the answers, as can be testified by several gentlemen who were present ; and they were again repeated two or three times the same evening with like results. This patient was operated on the previous day in presence of several professional and scientific gentlemen, when several answers were given different. More than one being operating on that occasion, and the manner and degree of touching the parts being different, might be the cause of the varied results. I am satisfied this patient knew nothing of phrenology ; and that she remembered nothing of what she said, or was done to her during these operations.

Case II. In this patient *friction* would *excite*, whilst *pressure* had *no effect* in calling forth manifestations. In this case, friction over the *ossa nasi* excited the desire for "something to smell," generally aromatic vinegar or Eau de Cologne ; over the chin, for something to eat ; over the tendon of the orbicularis, a slight tendency to laugh ; close upon the root of the nose, friction excited spectra, and round the orbit, in like manner, the same or similar spectra, differing in form, and hue, and combination, according to the

degree of *pressure* and *friction* applied ; over the organ
of tune, " to sing " ; over the back part of the base
of the head, expressed herself " very happy and comfortable " ; over combativeness and destructiveness, a
quarrelsome disposition, as manifested in word, look,
and action.   The other parts tried were less certain
or decided in this patient.

Case III. In this patient, friction excited the desire
" to waltz," when applied over the organ of tune,
and the desire " to walk," when applied to the organ
of wit, as mapped by the phrenologists, and in like
manner, " to sing," when veneration was the point
affected. Spectra, also, when the integuments were
rubbed against the margin of the orbits. Although
not corresponding with the phrenological charts, nor
with what occurred in the others, similar answers
were given when the same points were similarly
excited.

Case IV. When asked what she would like, when
manipulated as the others referred to, always answered,
" nothing at all," excepting over the most sensitive
parts of the cranium, when her answer was, " leeches to
my head."

Case V. Very much the same as the last.

I think the cases referred to support my position,
that the different results arise from the circumstance
of different parts of the integuments having different
degrees of sensibility, and thus exciting different ideas
in the mind when the same quality and intensity of
stimulus is applied to each part in succession. There
can be no doubt but the point under which the phrenologists have posited " combativeness and destructiveness," is the most highly sensitive of any part of the

cranium, and is always accompanied with symptoms of the patients feeling pain, and, as a matter of course, they will offer resistance, and attempt to free themselves from the offending cause ; and so of the rest, according to their respective impressibilities.

After the above remarks had been written, and my work sent to press, I met with the following most interesting case :—I was informed that a child, five years and a half old, who had been present when I exhibited the experiments on No. I. the same evening had proposed to operate on her nurse. The nurse had no objection to indulge the girl, never supposing any effect would take place. However, it appeared she speedily closed her eyes, when the child, imitating what she saw me do, placed a finger on her forehead, and asked what she would like, when the patient answered, "to dance"; on trying another point, the answer was, "to sing," and the two then had a song together, after which the juvenile experimenter roused the patient in the same manner she had seen me do.

The above circumstance being related to me the following day, I felt curious to ascertain whether there might not be some mistake, as there had been no third party present, and it depended entirely on the statement of the child, which induced me, when visiting the family the day after, to request permission to test the patient. This was readily granted ; and, to my astonishment, she manifested the phenomena in a degree far beyond any case I had tried ; indeed, she did so, with a degree of perfection which baffles description.

However frequently she was tried, the same expression of countenance, the same condition of the respiration, and similar postures of the body have been evinced, when the same points were pressed. Indeed, so highly susceptible was she, that, after a few trials, when I pointed a finger or glass rod over the part, without contact, similar manifestations resulted, only in a less rapid and more modified degree. I also found by trying No. II. *at an earlier stage*, that her susceptibility was almost equal to the present case. The following are a few of the more striking manifestations : pressure on the chin was followed by movement of the jaws, lips, and tongue, with the desire to eat ; on the lower part of the nose, "to smell"; insertion of tendon of the orbicularis, immoderate laughter, which, on being asked why she laughed, the answer indicated, it was from a sense of the ludicrous being excited ; over time, "to dance "; tune, "to sing," with pressure on the eye at same time, she did sing part of a song ; over the back of the head, No. I. she shuddered and retreated, under the impression that some one was about to take liberties with her, the same feeling of delicacy was also manifested *when any other part of the body was touched excepting the head and face ;* over apex of mastoid process, the desire to shake hands and be friendly ; the former, with No. IV., or adhesiveness, she would lean to, or clasp any one near her ; combativeness, the reverse ; destructiveness (it is very small), she was distressed from the notion that some one *was quarrelling with her ;* philoprogenitiveness, she always said, "hark, the poor child is crying !" secretiveness and caution, she would tell nothing ; benevolence, "*to travel*" ; veneration, she knelt down in the most solemn manner and prayed ; combined with hope an expression of ecstasy united with devotion ; over the eyebrows, spectra of all forms and colours, gay and

glowing, and when below the eye, the notion of the
sea, a ship, and people about to be drowned ; at a
farther trial, other manifestations came out equally, or
even more strikingly, according to the accuracy with
which the corresponding points were touched.   In particular, I must note what happened the first time I
touched imitation, which was entirely accidental, and
whilst, besides a relation of my own, there was present
a gentleman whose literary and scientific attainments,
and philosophic turn of mind, as well as high standing
in society, render him an ornament to our country.
Besides imitating every thing done or said in English
she imitated correctly French, Italian, Spanish, German,
Latin and Greek ; every word was spoken with the
utmost precision, and has been done several times
since before many professional and scientific gentlemen,
and ladies, who can bear testimony to the extraordinary
fidelity of pronunciation and emphasis.   I need scarcely
add, she could not do so when tested after being
awakened.   Many other patients I have since made do
the same, one a girl of only twelve years of age.

On Mr Hall's arrival in Manchester, previous to his
first lecture, I had the pleasure of seeing him at my
house, when I exhibited my experiments on this and
another patient, with which he seemed much gratified.
I also afforded him an opportunity of seeing them
again next day.   After I had them in the hypnotic
condition, I requested him to manipulate their heads,
which he did more minutely than I had done, and consequently brought out additional manifestations.   I was
on the alert to all he did and said, for I was determined
he should not have an opportunity of *prompting in any
way*, and most assuredly, by exciting acquisitiveness, he

very soon led the patient to steal a silver snuff-box
from a gentleman present, and it was most striking
the anxiety with which she returned it, on Mr Hall
removing the point of contact to conscientiousness,—
the movement of the arm was changed *instantly*, as if
automatically. I had never tried to excite either of
these two points. The other manifestations, which I
had previously seen developed, were precisely the same
under his manipulations as my own. I made several
attempts to excite the organ of benevolence, but without effect, until one day I accidentally placed my finger
so low as I should have considered to be the middle of
comparison, as marked on the busts, when she instantly
evinced the emotion in the most active manner, saying,
"poor creature, poor creature," and not content, as
many are, with mere *words* of compassion, she anxiously
presented us with all the money in her pocket. I
should not omit to add, that this patient is quite *unconscious* of all she or others do or say whilst in this
state, and did not know the location of a single organ.

It would only be an unnecessary waste of time to
detail at length all the cases I have had since of similar
manifestations, varying in degree according to the original constitution and habit of mind of each patient.
This variety is the most striking proof of the reality of
the phenomena. There are some patients who have a
sort of indistinct recollection of what had passed, as if
it had been a dream ; two in particular, who observed
they had an indistinct notion of what they were doing,
but felt irresistibly impelled, as it were, to do certain
things, even whilst they thought complying with the
predominant inclination would make them very ridiculous. This, I presume, referred to imitation and comi-

cality, and such like humorous faculties, which they displayed in a very remarkable degree. These patients
are highly respectable and intelligent, and manifested
the phenomena quite as prominently as the patient last
named, that of veneration and hope, also filial affection,
in a manner baffling description. Each knew only one
phrenological organ.

That I might be the better certified that all was
reality, I also got a relation of my own to submit to the
operation, and it was quite conclusive. She has a slight
recollection of *some* things which were said and done,
but of others seems quite oblivious.

I had also the opportunity of verifying the truthfulness of these various and interesting phenomena through
the kindness of Mrs Col. ———, who submitted to be
operated on by me in presence of her husband, as also
the Major; the Captain and Surgeon of the regiment;
a high dignitary of the church, and who is also an
eminently scientific gentleman; Mr Gardom, surgeon,
and other professional gentlemen; Mr Aspinal Turner,
and a number of others, both ladies and gentlemen. In
about three minutes after she was asleep, I placed two
fingers over the point named veneration, instantly the
aspect of her countenance changed; in a little she slowly,
and solemnly, and majestically arose from her chair,
advanced towards the table in the middle of the room,
and softly sank on her knees, and exhibited such a
picture of devout adoration as can never be forgotten
by any who had the gratification to witness it. She
was tested with a number of other faculties, when the
corresponding manifestations were equally striking and
characteristic. When awakened, this lady was quite
unconscious of all which had happened.

Here, then, we have the testimony of a lady of the highest respectability and intelligence, and energy of mind, corroborating, both in word and action, and look, the reality of the phenomena as exhibited by others, and that in the presence of most respectable and intelligent witnesses, who can bear testimony that there was nothing said or done to direct her in the important manifestations. This lady had been hypnotized by me once before, for a few minutes, at a private conversazione the week before, when she sat down fully convinced she could NOT be affected, but was soon made to acknowledge the power of hypnotism, and now she was a valuable evidence to the more novel investigation as to how far phrenological manifestations could be developed during hypnotism.[1] I have now realized these phenomena very prominently in forty-five patients, most of whom, I am quite certain, knew nothing of phrenology, some of them not even what the word meant; and the smallness of the points to which the contact must be made to elicit the manifestations correctly, especially the subdivisions by Mr Hall, is such as to render collusion most improbable. I was also careful to avoid prompting, by putting leading questions. I have also succeeded partially with others; and several of my friends have also been successful with a few other cases.

[1] A report having been circulated, no doubt with the view of neutralizing the interest attaching to the case, that this lady was a phrenologist, I called to inquire whether there was any ground for such a report. Mrs S. herself assured me it was quite erroneous, for it was a subject she had never paid any attention to, and one she was quite ignorant of. Wishing to be very circumstantially correct in the statement, she added, " I have understood the organ of music is somewhere about the forehead "; when requested to place her finger on the organ, she was quite wrong, so that *she did not know a single organ.* I mentioned this circumstance, in her presence, at another conversazione, when she most distinctly declared the facts here recorded, to be strictly correct.

I attended Mr Hall's public lectures, and the very first experiment he tried, February 24, 1843, convinced me, that the reason why I had not sooner obtained the manifestations more generally was, because I had allowed my patients to pass into the *supersentient* state before testing them. I was aware of the difference in the state of the circulation through the brain in the state in which my patients were, and what it must be in the state in which his were during his operations, and conjectured, that by trying my patients in *that* condition, I might get manifestations which I had failed to do at former trials ; and the very first cases I tried proved this conjecture to be correct.   For example, No. II. already referred to, exhibited a number of additional phenomena beautifully ; and Cases IV. and V. in like manner, came out beautifully.   From this single observation on Mr Hall's mode, or rather *time*, of operating, I have been enabled to arrive at a mode of operating in which, I believe, by putting patients into the hypnotic condition my own way, there will be no great difficulty in manifesting some of the phenomena in most cases.   There are some patients, however, who will evince them much more prominently than others, and the power of habit seems evident in most, being more readily operated on after a few trials.   Some, however, seem as perfect as possible at the first trial.

I have tried several private friends, on whose intelligence, honour, and integrity, I could rely, and also children, and have found the evidence so satisfactory, that I am quite certain as to the reality of the phenomena ; but as to my theoretical views, I wish them to be considered as mere conjectures, thrown out for the purpose of exciting others to think, and investigate this curious and most interesting and important subject.

I shall conclude this article by calling the attention of my readers to the coincidence which appears to subsist betwixt the phenomena now referred to, and the mode of exciting dreaming, in some patients, by whispering in their ears.   I shall illustrate that by reference to a case recorded in one of Dr Abercrombie's valuable works, on the authority of the late Dr Gregory.   It is that of the case of an officer in the expedition to Louisburgh in 1758.   His brother officers were in the habit of amusing themselves at his expense.   They could produce any kind of dream they chose, especially if done by one with whose voice he was familiar.   Thus, at one time, they conducted him through the whole process of a quarrel, ending in a duel ; and when it was supposed the parties met, a pistol was put in his hand, which he fired, and was awakened by its report.   On another occasion, being asleep on the locker of the cabin, he was made to believe that he had fallen overboard, and was told to save himself by swimming.   He imitated the art of swimming, when they told him to dive for his life, as a shark was pursuing him, which he attempted so energetically, that he threw himself from the locker, by which he bruised himself severely.   Again, after the landing of the army, he was found one day asleep in his tent, and apparently much annoyed with the noise of the cannonading then going on briskly.   He was made to believe he was engaged with the enemy, when he expressed much fear, and betrayed a wish to run away. They remonstrated against this act of cowardice, whilst they increased his alarm by imitating the groans of the wounded ; and when he inquired who was killed, which he often did, they named his particular friends. At last he was told that the man next him in the line had fallen, when he instantly sprang from his bed, rushed out of the tent, and was aroused from his sleep, and relieved from his fears, by falling over the tent ropes. It is added, that after these experiments, he had no distinct recollection of his dreams, but only a confused feeling of oppression or fatigue ; and used to say to his friends, that he was sure they had been playing him some tricks.

I shall add one illustration as to the probability of benefit accruing to society from this subject being prosecuted with zeal and due consideration. A highly scientific friend, who had honoured me with his presence at a private conversazione, called two days thereafter, and stated, that from reflecting on what I had said and exhibited the day before as to the mode of exciting certain points or functions of the brain through acting on certain points of the scalp and face, it appeared to him most reasonable to expect, that by applying treatment to *such* points, we might most readily afford relief to disorder of the corresponding internal organs. I told him I was so thoroughly convinced of that, as to have been induced to act accordingly ; and that the day before I had been visiting an insane patient, who entertained the horrible idea, that she must murder every body she knew, and then murder herself also ; that on placing my hand upon the organs of combativeness and destructiveness, in a few seconds, she gave a violent shudder, and seemed greatly excited, becoming perfectly furious. On examining these parts, I found the integuments quite red. I ordered leeches, and cold lotion afterwards, but next day she remained equally violent, and the pulse between 140 and 150, which it had been for some time, notwithstanding medicines had been given to depress it. I now made an incision an inch and a half long through the integuments, and down to the bone, and in twelve hours after found her much calmer, and the pulse down to 100, and it remained there for several days. There was no such loss of blood as could have acted constitutionally on the heart directly by the quantity effused. On again rising, Belladonna plasters were applied—these not having the desired effect, recourse was again had to scarification behind both ears, and with great success, as in a few days she was so calm as not to require the strait jacket, and for two months has been sullen but harmless.[1]

At another conversazione, the same gentleman requested me to excite philoprogenitiveness, which I did, and he then asked me to combine destructiveness along with it. I told him the faculty would not be developed, because the organ was so small in this patient as to make her always imagine some one was quarrelling with her. Still he wished me to try, which I did, and the result was that she immediately seemed distressed about some one being angry *with the children*. Two days after I was informed that the object of the request was to prove that *such would be the case*, as he had whispered to a professional gentleman present, before the answer was elicited, and no one else in the room knew this remark. Two days after, on a slip of paper handed to me by the same gentleman, he had noted, that if I would excite the same organs in another patient, whose destructiveness was more prominent, I would find she would be angry with the children, and

[1] See Appendix I., Note II.

wish to punish or send them away, and assuredly it proved so. He also added, that this is the combination of morbid excitement which he conjectured, and I think with great justice, is the cause of parents murdering their own children during a fit of insanity. An example of more acute, beautiful, and successful induction than this could scarcely be conceived possible ; and it is highly gratifying to know that the opinions of a gentleman of such talents and attainments coincides so much with my general views on this subject.

The doctrine propounded by the Rev. La Roy Sunderland, and Mr Spencer T. Hall, and others, seems to be this, that there is a separate organ in the brain for every mental faculty, emotion, propensity, desire, and action, mental or corporeal ; that every positive organ has also its negative organ proximate ; and that by certain manipulations during the mesmeric state, these organs may be stimulated into activity singly or combined, and thus caused to manifest the corresponding faculty by thought, word, and action. They do not deny the correctness of the outlines as given by former phrenologists ; on the contrary, they bear positive testimony to their general correctness. However, they subdivide each of the former faculties, which we may designate the pure faculties, into groups of distinct organs, for the specific manifestation of special faculties, the tendencies to which were naturally included in the simple or primitive general organ ; and they allege they can thus give such a special or characteristic direction to the feeling as to entitle it to be considered as the manifestation of a distinct organ or faculty.

It occurs to me, that this might be much simplified,

by considering, that on the central point of the *general* organ, we stimulate fasciculi of nerves connected with a general manifestation, for example, benevolence ; but that, as we approach the surrounding organs, we partially excite proximate faculties, from some of their corresponding peripheral sentient nerves co-mingling with those of the *other* faculty, and thus engender a mixed manifestation ; just as we find the intercourse between neighbouring countries modifies the national character which peculiarly belongs to each nation. Thus, in one direction, benevolence (by which I illustrate my position) will be blended with comparison, or excited through the influence of association respecting some one we have known, or from supposing what might be our feelings were we placed in such and such circumstances ; in another direction, it will be influenced more or less by the tendency to imitate the benevolent acts of others, and, as we approach veneration, it will partake more of a religious and moral obligation in reference to the Deity. If I am right in this conjecture, of course there will be every possible shade of manifestation as we approach nearer to the adjoining organ. I am not acquainted with the mapping of the head either by La Roy Sunderland or Mr Hall ; but, if the original compartments are to be so divided and subdivided, according to the mere varieties of manifestation during the hypnotic state, I feel assured, that each of their *subdivisions* may be again *divided*, as a shade of difference will be manifested by every possible change in the point of contact.

I had much pleasure in witnessing Mr Hall's experiments, and bore public testimony to the reality of the general phenomena. This I could have done from the mere circumstance of carefully watching the peculiar expression of countenance, and state of the respiration, induced by every move of the point of contact. The shades of difference were so minute that collusion was all but impossible. Moreover, I had personal experience of the reality of the leading phenomena in a number of my own patients, with parties who knew nothing of phrenology, and whose respectability and known character placed them above the possibility of being suspected as acting a part, either for the purpose of gratifying or deceiving others. Whilst I readily bear testimony to the reality of the phenomena, and that I saw nothing in Mr Hall to lead me to suppose he wished to deceive any one, it is due to the cause of truth for me to state, that the varieties which I observed in his phenomena and those occurring in my own patients, I consider were the mere results of the different manipulations used, and not of any such *special* influence as he and other animal magnetizers allege.

In reference to the phenomena which were designated "cross-magnetizing," and which appeared most distressing to the patients, as well as to the operator (fortunately no such effects have ever occurred in my patients), I think they may be explained thus : it seems probable, part may be the result of imagination, or an accidental circumstance exciting the opposing classes of muscles into action at the same time. This may also be caused by exciting two antagonist emotions, such as one requiring the energetic action of the muscles of inspiration, and the other the muscles of expiration, the consequence of which is, very speedily to throw the patient into a state of partial asphyxia ; and the result must be, a great difficulty in restoring the patient from the deleterious

influence of insufficiently decarbonized blood circulating
through the brain. Such I consider was the case with
the patient I saw create so much trouble to Mr Hall on
the evening of the 24th February, 1843, in the lecture-
room of the Athenæum, Manchester.

Having heard Mr Hall state, that patients who had
stolen any thing would always seek out the persons
from whom it had been taken, and restore it to them
after conscientiousness was excited, and that they
would find out the rightful owner whatever part of the
room he had removed to, I was curious to prove this.
My first object was to ascertain whether it was a fact,
which I very soon did with my own patients, and my
next object was to ascertain *by what means they accomplished this*, and I readily determined it was *by smell*
and *touch*. The first thing they did, on rousing conscientiousness, was to look thoughtful, then they began
sniffing, and traced out the parties robbed, and restored
it to them. When asked, what are you doing? the
answer was, "I am giving back something which I had
stolen." On being asked, how do you know the person?
(having gone to the opposite side of the room) the
answer was, "I smell them, or him." Every time the
experiment was tried, the result was the same, and the
answer the same, as was obvious to every one in the
room. Another patient did the same *when the sense of
smell was acute*, but when I tried the experiment with
the *sense of smell dulled*, the stolen article was *merely
laid down*, without giving it to the proper person.
There was thus both positive and negative proof of
exalted smell being the cause of them restoring to the
proper party ; and feeling directs as to place. I have
found this done with the same promptitude and cer-

N

tainty when six, eight, or twelve faculties had been roused and manifested before conscientiousness was excited. I have found this the same in all I have tried, only some will throw the articles down as if horror-struck.

The movement of the jaws also, and various other movements in imitation of the operator, I have ascertained arise from their remarkable power of hearing *extremely faint sounds*, and the most curious point is this, that they seem to have the power of discerning such *faint sounds*, when they seem not to be affected by *very loud* sounds. It is also the same with feeling. They will in some states be insensible to pricking, pinching, or maiming, but so highly sensible to a breath of air, or the tickling of a feather, that they may be instantly roused by the latter means, when the former would have no such influence. Probably this is the cause of the remarkable effects of a current of air, its rousing cutaneous sensibility, directing the nervous influence to the skin, and withdrawing it from rigid muscles, thus reducing the cataleptiform state, and permitting the blood and *vis nervosa* to flow in their usual manner. The latter change being induced gradually, may probably be the cause of the feeling which is described as that of needles and pins running into the extremities, and producing twitching, when gently pressing on the extremity with the finger, etc., as already noticed.

In concluding this chapter, I am well aware the statements it contains must appear startling, and almost beyond belief, to many of my readers. Some may be disposed to think I have been deceived ; and because many of the manifestations *might* be simulated, I know it has been alleged, that the patients of those

who have been exhibited publicly, were either deceiving
the operator, or that both patients and operators were
engaged in a shameful system of collusion. In respect
to my own patients, I have endeavoured to take every
possible precaution that they should *not* deceive me,
and with this view have invited the most sceptical
persons I know, both in the profession and out of it,
to have it *rigorously* tested, and the result has been my
entire conviction as to the reality of the phenomena in
my *own* patients, and I am ready to believe others to
be as candid as myself. Because much *might* be simulated, and parties have been avowedly trained and exhibited to prove the dexterity of teachers and pupils in
a system of avowed collusion, that it might thereby be
inferred the patients exhibited by other lecturers were
impostors, is a most illogical mode of deciding such a
question. There ought to be positive proof of the justice
of such imputation, before so assailing any one, when
there is so much proof to the contrary, as has been
furnished by the concurrent testimony of so many ex-
perimenters who have met with such susceptible subjects.
Surely it would not be fair to infer, that because some
are trained as dexterous thieves, there can, therefore, be
no such thing as an honest man in the world?

The question to be decided here is not what patients
can be trained to do in violation to nature's laws ; that
is, by giving them some stronger motive of action, by
artificial means, than the impulse arising from natural
feeling. What might be achieved in this way I know
not, as I have not tried such experiments connected
with this branch of the subject. It is well known,
however, that so long ago as December, 1841, I particularly pointed out the remarkable docility of patients

during Hypnotism, which made them most anxious to comply with every proper request or supposed wish of others. I have, therefore, no more doubt that they might be trained to manifest, during Hypnotism, opposite tendencies, in accordance with conventional arrangements, than that during their waking moments they could be taught to do so, and thus call black white and white black, night day and day night, and such like, in respect to every custom, word, or action. The proper question to be determined seems to me to be this,—Can the passions, and emotions, and intellectual faculties, be excited during Hypnotism simply by contact or friction over certain sympathetic points of the head and face, without previous knowledge of phrenology, training, or whispering, or such leading questions as must naturally excite in the mind such passions, emotions, or mental and bodily manifestations? My own experience seems to warrant me to answer in the affirmative, and I shall give a few additional cases in illustration of the data from which I have come to this conclusion.

Two patients, healthy, strong servant girls, entirely ignorant of phrenology, neither of whom had ever seen an experiment, and one was so sceptical, as to wish to try and convince me *she could not be hypnotized at all*, were operated on separately. At first trial, I succeeded in hypnotizing both, and in developing a great number of the leading organs, such as the desire to eat, benevolence, friendship, pity, attachment, self-esteem, love of approbation, imitation (when they readily spoke five languages correctly), stealing under acquisitiveness, and under conscientiousness restored to the proper person and place what was stolen; eventuality most remarkable; this was tried twice or thrice in each, when they could tell correctly the events of the previous day while the organ was excited, whereas they could not tell a single circumstance before it was stimulated; and a number of others, such as forms, figures, and colours, by exciting the corresponding points. These experiments were tried before several friends, who were astonished with the result, several of the most remarkable manifestations being evinced without a single word being spoken by any one. They were not tried at the same time, and neither saw nor knew of the other.

Mr T., a gentleman of 45 years of age, who was ignorant of phrenology, and had never seen a phreno-hypnotic experiment, was hypnotized without expecting any experiment of the kind to be tried. On touching "benevolence" the manifestation was so powerful as to compel me quickly to desist; "self-esteem," very decided; "ideality," very decided, combined with "tune and language," he sang when the latter were pressed on, but instantly stopped when the pressure was removed, and resumed as readily on renewing the contact, exactly at the same note and word where he left off. Also the usual spectra when the region of the orbit was pressed on. When aroused, he was quite unconscious of all which had happened. He has been tried three times, with the same results, only that additional manifestations came out. His friends, who were present, can testify he had no signal given to lead him to do so. His wife, also, who had never seen any thing of the kind before, was operated on, when a great many manifestations came out most decidedly. Their daughter, who had seen nothing of this, was now called into the room, and operated on, and exhibited a great many manifestations, and all this by the mere effect of pressure and gentle friction on the integuments. None of the three remembered any thing of what had happened.

W. T., a boy, had been magnetized, and exhibited a few manifestations. He was again tried in public, but without success. I was requested to try him, when a number of manifestations came out at once beautifully —under benevolence, he took off his coat to give to some distressed person, and after a number of other manifestations had been educed, on being awakened he seemed very much surprised to find his coat off.

John W., 22 years of age, had been magnetized publicly, with the hope of eliciting the phrenological manifestations, but he became so stolid that it was quite a failure. I was afterwards requested to try him, in my way, in the presence of a number of gentlemen, when I at once succeeded in exciting several ; pity was so characteristic, that there could be no difficulty or doubt on the subject, as it was not only exhibited by his features and sobbing, but by the tears which ran over his face in torrents. On trying to excite imitation, on the right side, no effect was produced, which I suspected to be the result of an injury he had sustained, which had destroyed the integuments, and also caused exfoliation of the outer table of the skull. I therefore tried in the opposite side of the cranium, when the faculty was manifested beautifully. This seems a good corroboration of my theory, that it arises from the peculiar condition of the nerves of the scalp. On further trials many more came out without any cause beyond the simple excitation of the integuments by pressure and friction. Not only may such general manifestations be thus

excited, but, what is far more curious, by exciting
antagonist points in the *opposite hemispheres* of the
brain, the patients may be made to exhibit correspondingly opposite feelings in the different sides of
the body. If the antagonist faculties are excited on
the *same* side, there will be exhibited only the
stronger of the two. These "opposite influences on
the two sides," as Dr Elliotson has well remarked,
"are the most astonishing and beautiful experiments
that all physiology affords"; and are also the most
beautiful examples of the correctness of Mr Mayo's
fifteenth aphorism, at page 28 of his Nervous System
and its Functions, where he says, "Each lateral half
of a vertebral animal is separately vitalized. Or the
preservation of consciousness in one half is independent of its preservation in the other." It is true
that vivisections have proved this, but neither so
beautifully nor humanely as in the experiments I
now refer to, and those already recorded at page 136
of this treatise. Miss S., a lady who had never seen
a phreno-hypnotic experiment, and knew nothing
of phrenology, exhibited at first trial a great number
of the leading manifestations, and at a second and
third, these opposite ones in a remarkable manner.
Under friendship and adhesiveness, she embraced
a female friend in the most affectionate manner, and
on destructiveness being excited *on the opposite side of
the head,* she rushed forward with great impetuosity
to repel some imaginary adversary, whilst, with her
other arm and hand, she contrived to shield her
friend. Had I not laid hold of her, she would most
certainly have rushed through the window. On being
roused she was quite oblivious of all she had done.

Mrs C., another equally ignorant of the subject, displayed the same phenomena. The effect of music in exciting to ecstasy, elegance of movement, and graceful dancing, was most remarkable. Remembered nothing. Miss ———, entirely ignorant of the subject, and had never seen an experiment of the kind, and expected only to be *attempted* to be hypnotized, but whilst she wished to be tried, she had expressed to the friend who introduced her, that she could not be made to sleep. She exhibited veneration solemnly, with hope, glowing devotion, and with ideality and language, overwhelming ecstasy, expressing her happiness and prospect of entering into heaven ; " self-esteem," the most conceited prude ; " firmness " most decided ; " adhesiveness and friendship," and this in one side, and " combativeness and destructiveness " on the other at pleasure ; " imitation " in perfection, speaking correctly every language tried ; " benevolence " extremely marked, to the effusion of tears ; " acquisitiveness, conscientiousness, eventuality, the desire to eat, to smell, spectra," etc. etc. She was quite unconscious of all that had happened, and the friend who brought her to me knows she had no prompting. She has been tried once since with the same results.

Some parties, who were excellent critics, after seeing the latter and two others operated on, and expressing their utter astonishment with the accurate and natural manner in which every passion and emotion was manifested, expressed a strong desire to see some one operated on for the first time. I offered to operate on any of three young ladies whom they had introduced to me that afternoon, and whom I had not known pre-

viously ; indeed, one was a stranger in town, from the
south of England, who knew nothing of hypnotism or
phrenology, and had no faith in either, notwithstanding
what she had just seen. She, Miss S., sat down an
entire sceptic, but in a few minutes she was not only
most decidedly 'hypnotized, but also one of the most
beautiful and decided examples which could possibly
have been met with of the phrenological sway during
hypnotism, simply by stimulating the nerves of the
scalp and face. The moment " veneration " was
touched, her features assumed the peculiar expression
of that feeling, the hands were clasped, she sank on
her knees in the attitude of the most devout adoration ; combined with " hope," the features were illumin-
ated, and beamed with a feeling of ecstasy, the hands
being unclasped and moved about in the utmost
delight ; and when " ideality" was added, the ecstasy
was so extreme as scarcely to be supportable. On
changing the point of contact to "firmness," she instantly arose, and stood with an attitude of defiance ;
" self-esteem," flounced about with the utmost self-
importance ; the "love of approbation " was painted
to the greatest perfection ; "imitation," imitated accurately every thing done or spoken in any language ;
" friendship and adhesiveness," clasped hold of me ;
and by stimulating "combativeness" on the opposite
side of the head, along with the other, she struck
out with the arm of the side on which combativeness
had been touched, but held me fast, as if to protect
me, with the other. Under "benevolence," she seemed
much affected, and distributed her property to the
imaginary distressed objects her fancy had painted ;
under "acquisitiveness" she stole, and under "con-

scientiousness" she restored ; "tune," the desire for
music, and sang beautifully, a waltz being played, she
danced with a grace and elegance surpassing all which
any of us ever witnessed. Eventuality was also most
remarkable ; the desire to eat, to smell, was also excited ; also form, figures, colours, etc. ; philoprogenitiveness admirable. All this was done at first trial,
with an entire stranger, and the lady's immediate
friends, as well as others present, can bear testimony
that there was not the slightest prompting either by one
or other, and when awakened she was quite unconscious
of all which had happened. This lady has been twice
operated on since, when all these manifestations, and
many others, were exhibited in the most perfect manner,
as can be certified by Sir Thomas Arbuthnot, Major
Wilbraham, Colonel Wemyss, the Rev. Mr P., and
another high dignitary of the church, and the patient's
family and friends ; and that when under "number"
she wrote down a sum, and under "constructiveness
and ideality," she drew a very good sketch of a cottage,
putting in doors and windows correctly. The uncle of
the latter subject was so much astonished and gratified
with what he had seen, that he begged I would try one
of his daughters. I hypnotized the eldest, and all the
manifestations came out quite as decidedly as in her
cousin. Under "adhesiveness and friendship," she
clasped me, and on stimulating the organ of "combativeness" on the opposite side of the head, with the
arm of that side she struck two gentlemen (whom she
imagined were about to attack me), in such a manner
as nearly laid one on the floor, whilst with the other arm
she held me in the most friendly manner. Under
"benevolence," she seemed quite overwhelmed with

compassion; "acquisitiveness," stole greedily all she could lay her hands on, which was retained whilst I excited many other manifestations, but the moment my fingers touched "conscientiousness," she threw all she had stolen on the floor, as if horror-stricken, and burst into a flood of tears; on being asked, Why do you cry, she said, with the utmost agony, "I have done what was wrong, I have done what was wrong." I now excited "imitation and ideality," and had her laughing and dancing in an instant. On exciting form and ideality, she seemed alarmed, and when asked what she saw, she answered, "The D—l." What colour is he? "Black." On pressing the eyebrow, and repeating the question, the answer was, "red," and the whole body instantly became rigid, and the face the most complete picture of horror which could be imagined. "Destructiveness," which is largely developed, being touched, she struck her father such a blow on the chest as nearly laid him on the floor. Had I not endeavoured to restrain her, he must have sustained serious injury. Having now excited veneration, hope, ideality, and language, we had the most striking example imaginable of extreme ecstasy, and on being aroused, she was quite unconscious of all that had happened, excepting that she had heard music, and had been dancing. Her philoprogenitiveness was admirable.[1]

[1] There were a dozen present on this occasion, of whom Mr Vandenhoff was one. Being well known as an accomplished artist, I requested him to watch all he saw with the most critical attention, and to tell me whether the passions were painted naturally or the contrary. After witnessing the first case with evident delight and surprise, he made the following observation,—"*If* this is *acting*, it is the *most perfect acting I have ever seen.* In acting, we aim at being natural, but there is generally some point in which we fail; but here I see nature's language in *every* point." Similar expressions followed, in what was seen in the next two cases, and when

At a conversazione a few days after, in the presence
of Lady S., Sir Thomas Arbuthnot, Colonel Arbuthnot,
Major Wilbraham, John Frederick Foster, Esq., Chairman of the Quarter Sessions, D. Maude, Esq., stipendiary
magistrate, and many others, both gentlemen and ladies,
after exhibiting the phenomena on those who had been
previously tested, there was a wish expressed to see
some one operated on *for the first time.* I offered to
try any one present, and a lady at length consented,
whom I never saw before that day, nor since. She
exhibited all the usual phenomena very decidedly.
Under " acquisitiveness," she stole two handkerchiefs
from ladies, and a ring from Mr Foster's finger. After
several manifestations had been exhibited, the moment
I touched " conscientiousness," she seemed distressed,
and set off and searched out the proper parties to whom
to restore the respective articles. They had changed
places, but she found them out, and gave back the
handkerchiefs to their owners, and also put the ring on

he witnessed the effects on the two ladies, whose cases have just been
recorded, he confessed himself so overpowered, as to be scarcely capable
of expressing his feelings of delight and astonishment, but said he should
write me on the subject. The following is part of a letter I received from
him two days after,—" I thank you for your kind invitation to witness a
repetition of those experiments which so much delighted me on Saturday
last, and with the result of which I was no less gratified than astonished.
Never have I seen nature manifesting herself more distinctly—never *so
beautifully*, as in the course of the exhibition on that evening. I believe
you know I was a decided sceptic in the mesmeric influence—and I was
something more in relation to its phrenological sway—of which the manifestations while under its mysterious influence, by the two young ladies of
my own immediate acquaintance, who had not, who could not have had,
any knowledge of the subject prior to their experience on that evening,
have perfectly convinced me by their truthfulness. I may take a farther
opportunity to dilate more fully upon this interesting and wonderful discovery, the beneficial results of which cannot yet be appreciated, because
we know not to what extent they may be carried out."

the very finger of Mr Foster from which she had taken it. She was a strict methodist, who had never danced in her life, and who, if awake, would have considered it a sin to dance. However, under the excitement of suitable music, she cut a very good figure at waltzing. When awakened, she remembered nothing of all which had happened.

Miss L., a lady of twenty-one years of age, very accomplished, and with great energy of mind, braved me to try to hypnotize her. She felt assured I could not do so. However, she was very soon under the influence, and gave twenty manifestations in the most decided manner. Under friendship and adhesiveness, and destructiveness on the opposite side, she protected me, and struck her own mother. She knew only one organ, and was inclined to scoff at Hypnotism, and still more so at phreno-hypnotism. Under form and ideality she wrote very nicely, without the use of her eyes, but by no means equal to what she does when awake. When awakened she seemed surprised when told what had happened. She remembered me touching her head, wondered what I was doing it for, said she felt different impulses arise when I was manipulating different parts, but did not know why, nor could she remember what she had done.

A married lady, Mrs E., and the mother of a family, would not believe any one could be so affected. After seeing one patient done, she still felt assured *she*, at least, could not be so operated on. I desired her to try, and she at once exhibited upwards of twenty manifestations in the most distinct manner, some of them very strikingly. Under benevolence she shed tears, drew out her purse, and gave half-a-crown " to the poor creatures."

She also exhibited the opposite tendencies at the same
time, as already described.

Miss R., a young lady of 22 years of age, very well
educated, and intelligent, wished to be tried, because
she was decidedly sceptical.   It so happened that every
manifestation tried came out beautifully and prominently,
although, when aroused, she admitted she remembered
every thing she had done, and added, that she had
resisted to the utmost of her power doing any thing, but
felt irresistible impulses come over her to act in the way
she did, as I touched certain points, but *why* it was she
could not tell.   Declared it was not from any association with what ought to be the case, as she was ignorant
of the organs, but added, that she first felt a drawing in
the muscles of the face, and then the breathing became
affected, and with this the peculiar impulse followed.
On another occasion, with the eyes bandaged, she had
a pencil put in her right hand, when a number of organs
were excited, but she showed no evidence of any desire
to use the pencil till "constructiveness and ideality"
were excited.   The moment this was done, however,
she scrambled till she got some paper, and began
drawing, and made a very tolerable profile.   When
"acquisitiveness" was excited, she stole a ring off Mr
Foster's finger, who, while I was exciting various manifestations, left the room.   The moment I touched "conscientiousness," she set off in search of Mr Foster,
walked round the room the very way he went, then left
that room, crossed the lobby into the front parlour, and
having made a gyration in this room, she came out and
went into a back parlour, where she found Mr Foster,
and put the ring on the very finger from whence she
took it.   She evidently traced him through the air by

smell, as she followed the exact track he had taken, for he had first gone into the front parlour. Had it been by clairvoyance, she of course ought to have gone to him direct, and by the shortest way. Such facts are almost past belief, but here they are as they happened, and there could not have been more competent individuals, than those present, to detect any mistake or deception, namely, Mr Foster, Mr Brandt, and Mr Lloyd, barristers ; Mr Langton, Mr Bagshaw, Mr Schwabe, and many others, both gentlemen and ladies.

Miss W., a very intelligent lady, who knew nothing of phrenology, and had never seen a phreno-hypnotic experiment, was operated on. On " benevolence " being excited, she seemed very distressed, and when asked what she was thinking of, said it was of a poor man who had lost his saw and hammer, that he had no money to purchase others with, and his children were starving. Under " veneration and ideality," wished to die, to go to heaven ; under combativeness, first looked very angry, then jumped up and gave a blow, which upset the candlestick. On " destructiveness " being excited (after she had exhibited several other organs), she shook her fist, then started on her feet, looked furious, and sprang across the room, her arm at full length, similar to a person fencing, and seized hold of a young lady's hand, and nearly transfixed it with her nails.

Mr Walker, twenty-two years of age, after passing into the hypnotic state, showed no symptoms of susceptibility for some time, but at length he did so in the most perfect manner ; namely, benevolence, veneration, firmness, self-esteem, combativeness, destructiveness, acquisitiveness, caution, conscientiousness, imitation in perfection, pity, benevolence with the one side, and destructiveness on the other, eventuality, smell, form, colours, number, ideality, etc. This gentleman has seen busts and phreno-hypnotic experiments also, but, excepting two or three, would be puzzled to point out any of the organs correctly when awake. He remembered nothing of what had passed.

Being desirous of ascertaining whether he might not, during hypnotism, remember the organs better than whilst awake, and thus be led to give the manifestations in the manner he did, I tried the following experiment. I explained my intentions to the friends who were to be present, but he was entirely ignorant of them. He had never seen or heard of such experiment having been tried. When I considered him in the proper condition, I requested him to place the point of a finger on different organs, but it was remarkable that he was wrong *in every* instance, even with respect to the few he knew when awake. Another most interesting fact was discovered, that whilst his mind was directed to the organ I had named, the *true* manifestation of *the point touched* came out *in every instance.* Thus, when requested to point out ideality, he placed the finger over "veneration," and immediately indicated that feeling. When asked what he was thinking about? "I did not go to church yesterday." What of that? "It was wrong." When he accidentally pressed on benevolence, the feeling was manifested; firmness in like manner; self-esteem in a powerful degree. On evincing symptoms of uneasiness, I asked what he was thinking of? he replied, "something hurts my head." The fact was, his arm had become cataleptiform, and the points of the fingers were pressed so strongly against the scalp as to be the cause of complaint, but he had no idea of that.

His hand having rested on philoprogenitiveness, he began to hush and rock on his chair as if nursing a baby, his motion became more and more violent till I judged it necessary to put a stop to it, by removing his hand. However, I found his arm and neck had both become so rigid, that they were too firmly fixed to permit of being separated by mechanical force, but so soon as this was reduced, by blowing on them, the peculiar manifestation ceased. Every point pressed on by him showed the same tendency to excite its peculiar manifestation. I am quite certain this gentleman acted a candid part, and could not be induced to do otherwise by any one.

Another most interesting fact connected with the latter case, was the circumstance of his having hypnotized himself, excited the different manifestations as stated ; and on being requested to rub his eyes, he did so, and thus roused himself from the hypnotic condition. I have tried similar experiments with many other patients, and, with the exception of two, each of whom hit upon one organ, have found none of them could point accurately to the organ named, but in every instance the usual indication of the peculiar organ touched came out. None of these subjects remembered any thing of what had happened. Here, then, we have decided proof, that all the phenomena of hypnotizing, exciting the phrenological manifestations, and rousing to the waking condition, may be accomplished by the personal acts of the patient on himself, as the only influence required to excite him to the necessary movement might be conveyed by an automaton.

A few days ago, one of these patients, who knows no foreign language, when imitation and tune were excited, followed correctly both the music and words of Italian,

French, and German songs, which she never heard till they were played and sung by the wife of a learned barrister, who was also present himself, and who, with the Rev. Mr F. and his lady, can bear testimony to the great accuracy of her performance.   Such is the power of Hypnotism.

Besides the twenty-five cases here briefly recorded, I have had many more exhibiting the phenomena in the same decided manner, simply by exciting the sympathetic points by contact. If I am to believe the evidence of my senses, therefore, *in any thing*, I cannot see how I can doubt that some relation subsists between certain points of the cranium, and the mental manifestations, which are excited by acting on them during Hypnotism.   I believe there are very few physiological phenomena which can be more clearly demonstrated, especially at such an early stage of their investigation. Were it not that I consider it would only be an unnecessary waste of time to prosecute the investigation farther, after the number of most unequivocal cases which have been met with by myself, as well as by other experimentalists here and elsewhere, I feel convinced I might soon increase the number of my own cases to any extent I chose.

With all intelligent and honest experimentalists I anticipate similar results to what happened with Mr Ebbage, an intelligent surgeon at Leamington. He had been a determined sceptic, and had much annoyed several of our mutual friends by his strong expressions to that effect. However, whilst on a visit at Manchester lately, at our first interview, I made a convert of him by offering to exhibit the phenomena in his own wife, who had never been so operated on, or even tried the experiment. She soon became decidedly hypnotized, and also exhibited several phrenological manifestations most distinctly. A servant of the family was now called into the room, who had seen no operation of the kind, and did not know what was to be done. She also became decidedly hypnotized, and exhibited several phrenological manifestations most distinctly. Mr E. now admitted that rational scepticism could not resist such conclusive evidence ; and having seen another case or two at my house, of remarkably susceptible subjects, with instructions from me how to operate, he promised to prosecute the inquiry on his return home.

In a letter to me, dated 1st May, 1843, he writes that he had tried the experiments with several ; that in some he was unsuccessful, while " in others a perfect state of sleep and unconsciousness was produced at different periods, varying from two to ten minutes. In the case of one lady, who had never seen any thing of the kind before, and, I may add, had not even heard it spoken of as connected with any phrenological developments, the most marked effects were soon produced, resembling very strongly the case you showed me when I was at your house." He farther adds the following judicious remarks :—" I must say the peculiar development shown by the influence of this sleep, if closely and scrutinizingly watched, must open to the mind of any thinking man a wide expanse for speculation as to the truly mysterious means by which the effects of sensation and emotion can be produced."

The above is a good illustration of what may be done, even by a determined but honest sceptic. Mr E. had only two interviews with me ; and if any one should be less successful in his attempts, it behoves him to inquire whether his failures are not to be attributed to his unskilful or uncandid performance of the experiments, rather than to inefficiency of the method recommended.

As to those who will not believe the testimony of others without seeing the experiments tried before themselves, on fresh patients, I beg to remark, that the best plan is for them to try patients *fairly* themselves, and they must soon be convinced ; only they must be careful to take them *at the proper time*, otherwise they may fail as I did myself at first.

The following is the mode of operating :—Put the patient into the hypnotic condition in the usual way, extend his arms for a minute or two, then replace them gently on his lap, and allow him to remain perfectly quiet for a few minutes. Let the points of one or two fingers be now placed on the central point of any of his best developed organs, and press it very gently ; if no change of countenance or bodily movement is evinced, use gentle friction, and then in a soft voice ask what he is thinking of, what he would like, or wish to do, or what he sees, as the function of the organ may indicate ; and repeat the questions and the pressure, or contact, or friction, over the organ till an answer is elicited. If very stolid, gentle pressure on the eyeballs may be necessary to induce him to speak. If the skin is too sensitive, he may awake, in which case try again, *waiting a little longer ;* if too stolid, try again, beginning the manipulalations *sooner*.

The operations should be tried again and again with the same patient, varying the time of beginning the manipulations, as it is impossible to tell, *à priori*, the exact moment they should be commenced; and many of the best cases have only succeeded partially, or not at all, at a first or second trial. When this point has been hit upon, however, there will be little difficulty in getting out additional manifestations, and this will be still more evident at each succeeding trial.

Whispering or talking should be carefully avoided by all present, so as to leave nature to manifest herself in her own way, influenced only by the stimulus conveyed through the nerves of touch exciting to automatic muscular action. We all know that during common sleep a person unconsciously changes from an *un*comfortable position to one which is agreeable. This is a sort of instinctive action, and, as already explained, I think it highly probable, that by thus calling into action muscles which are naturally so exerted in manifesting any given emotion or propensity, they may, by reflection, thereby rouse that portion of the brain, the activity of which usually excites the motion. In this case there would be a sort of inversion of the ordinary sequence, what is naturally the consequence becoming the cause of cerebral and mental excitation. The following hypothesis will illustrate my meaning. It is easy to imagine, that putting a pen or pencil into the hand might excite in the mind the idea of writing or drawing; or that stimulating the gastrochnemius, which raises us on our toes, might naturally enough suggest to the mind the idea of dancing, without any other suggestion to that effect than what arises from the attitude and activity of the muscles naturally and necessarily brought into play whilst exercising such functions. However, I would very much doubt the probability of stimulating the muscles of the leg ex-

citing the idea of writing, or that placing a pen or
pencil in the hand would excite the idea of dancing,
without previous concert and arrangement to that effect.
It is upon the same principle, as I imagine, that, during
the dreamy state of hypnotism, by stimulating the sterno-
mastoid muscle, which causes an inclination of the head,
the idea of friendship and shaking of hands is excited
in the mind, and when the trapizus is excited at
same time, the greater lateral inclination of the head
manifests still greater attachment, or "adhesiveness."
Philoprogenitiveness, by calling into action the recti
and occipito frontalis muscles, gives the rocking motion,
and hence the idea of nursing, etc.; pressure on the
vertex, by calling into action all the muscles requisite
to sustain the body in the erect position, excites the
idea of unyielding firmness; veneration and benevolence, from giving the tendency to stoop and suppress
the breathing, thus create the corresponding feelings.
By exciting the muscles of mastication into action, the
idea of eating and drinking is roused, and the same
may arise from pressing between the chin and under
lip, which first excites a flow of saliva, and this again
the motion of the tongue and jaws, with an inclination
to swallow. In like manner, gently pressing the tip of
the nose, by exciting inspiration, creates the desire for
something to smell at; if the point of contact is the
cheek, under the orbits, over the exit of the *infra-*
orbital branch of the fifth pair, the breathing becomes
suppressed, and depressing emotions are excited; whereas, *above* the orbit, so as to stimulate the *supra*-orbital
branch of the fifth pair, generally the reverse manifestations are evinced.

Those familiar with Professor Weber's experiments,

know that each of those points differs from the other in its degree of sensibility. It is remarkable that the point marked "eventuality" (and which I have strong grounds for believing is the chief seat of memory), is in the centre of the forehead, which is one of the most sensitive parts of the scalp, and where pressure applied necessarily excites the corresponding points in *both* hemispheres of the brain at same time.[1]

There seems, in fact, to be less matter of wonder in this discovery than some lately brought forward in other departments of physical science; for example, who would have believed, till it was proved, that by looking into a camera-obscura for a few minutes, or even seconds, he might have his likeness accurately and indelibly transferred to a plate of metal? or the still more recent discovery of Professor Moser, that such impressions as he referred to could be effected in the dark?

I shall conclude this article by a quotation, from Dr Abercrombie, on the value of testimony. He observes,—"A very small portion of our knowledge of external things is obtained through our own senses ; by far the greater part is procured through other men, and this is received by us on the evidence of testimony. While an unbounded credulity is the part of a weak mind, which never thinks nor reasons at all, an unlimited scepticism is the part of a *contracted* mind, which reasons upon imperfect data, or makes its own knowledge and extent of observation the standard and test of probability." *On the Intellectual Powers*, pp. 71, 72.

[1] See Appendix I., Note 12.

CHAPTER VII

BEFORE concluding the first part of this treatise, I shall make a short resumé of what I consider the points made out by what has been advanced. 1st, That the effect of a continued fixation of the mental and visual eye in the manner, and with the concomitant circumstances pointed out, is to throw the nervous system into a new condition, accompanied with a state of somnolence, and a tendency, according to the mode of management, of exciting a variety of phenomena, very different from those we obtain either in ordinary sleep, or during the waking condition. 2nd, That there is at first a state of high excitement of *all* the organs of special sense, sight excepted, and a great increase of muscular power ; and that the senses afterwards become torpid in a much greater degree than what occurs in natural sleep. 3rd, That in this condition we have the power of directing or concentrating nervous energy, raising or depressing it in a remarkable degree, at will, locally or generally. 4th, That in this state, we have the power of exciting or depressing the force and frequency of the heart's action, and the state of the circulation, locally or generally, in a surprising degree. 5th, That whilst in this peculiar condition, we have the power of regulating and controlling muscular tone and energy in a remarkable manner and degree. 6th, That we also thus acquire a power of producing rapid and important changes in the

state of the capillary circulation, and of the whole of
the secretions and excretions of the body, as proved
by the application of chemical tests. 7th, That this
power can be beneficially directed to the cure of a
variety of diseases which were most intractable, or
altogether incurable, by ordinary treatment. 8th, That
this agency may be rendered available in moderating
or entirely preventing, the pain incident to patients
whilst undergoing surgical operations. 9th, That during
hypnotism, by manipulating the cranium and face, we
can excite certain mental and bodily manifestations,
according to the parts touched.

I have obtained analogous results with so many
patients, as to make me quite certain of the *reality of
the phenomena* referred to, and to warrant me, as I
think, to draw these inferences. Many of the phenomena are of such a nature as to admit of physical and
chemical proof, in respect to which, the patients cannot possibly deceive us ; and as regards those phenomena where they *might* do so, I have had the assurance
of so many patients, on whose veracity I can implicitly
rely, proving the same facts, that there remains not
the slightest room for *me* to doubt the correctness of
these statements. I have been equally anxious to
avoid being myself misled, as I should be not to mislead others ; and I would recommend those who have
not had an opportunity of watching such phenomena,
in the most critical manner, or who have not entered on
the investigation with candid minds, to suspend their
opinions until they have had such opportunity. I
have no hesitation in saying it is most improbable
that any man should form a just estimate in this matter from *mere reading* or *hearsay evidence*, and equally

so if he does not approach it with a mind open to honest and fair investigation. The subject itself is so very subtle in its manifestations, so very different from all we are accustomed to meet with in the *ordinary* condition, that, with the utmost candour and openness for receiving the truth, and the whole truth, it will be found extremely perplexing to follow it out in many of its bearings. How then can it be expected any one should prosecute the enquiry successfully who enters on it with his mind blinded by indomitable prejudice ?[1]

[1] It would perhaps be difficult to adduce a stronger proof of the extent to which prejudice may overcloud the brightest intellects, and render them incompetent to do justice to the subject they would investigate, than that which was presented at a late meeting of the Medico-Chirurgical Society of London, when a debate took place after the reading of Mr Ward's case of amputation of the leg during mesmeric sleep. As I am not an animal magnetizer, nor personally acquainted with any of the parties referred to, any remarks I am about to make are of course uninfluenced either by pique or prejudice.

The operation referred to was said to have taken place in a public hospital ; in the presence of medical, and also non-medical witnesses. The patient is alleged to have exhibited no manifestation of feeling pain, as far as his countenance could be taken as a correct index, and there was no movement of the limbs or body ; and after the operation he is said to have declared that he did not feel any *pain*, but had heard " a grunching," which it has been inferred was the noise of the sawing of the bone ; and it was also admitted he had groaned during the time he was under the operation.

How was this announcement met ? First, it was questioned whether the man was not a person of *little* or *no feeling* at *any* time, because *other* patients had been known, whilst wide awake, who were very insensible to pain. But had not the patient, in this case, been declared to have been suffering so much pain from his knee, that he had been unable to sleep, and that his health was so much impaired by his suffering as to render amputation of the leg indispensable ? Nay, had it not been set forth, that the pain of his leg had been greatly diminished, and his sleep restored, and his health greatly improved, after he was mesmerized, *preparatory to the operation*, which he had consented to undergo whilst in that state ; and yet, that after he had been asleep, and considered in a fit state for being operated on, the mere movement of the joint, whilst drawing him to the edge of the bed, was followed by *so much pain as to awake him.* Was this any proof of his being a person devoid of feeling?

Then it is held, that as he *heard*, as it is presumed he did, the *sawing*

As to the proximate cause of the phenomena,[1] I believe the best plan in the present state of our knowledge, is to go on accumulating facts, and their appli-

of *the bone*, he *must* have *felt* the *cutting of the skin and soft parts*. It is thus assumed that it is *impossible* for a person to *hear*, and be in the state *not to feel inflictions on the limbs at same time*. It is well known, however, that disease of the trunks of the sentient nerves, or of the spinal cord, may induce such a state, independently of any lesion of the brain. But then, say others, had he *not felt* when the *principal nerve was irritated*, the *other* leg *must* have been convulsed. This is assuming, that the speakers *fully* knew *every* law which *has* been known, or *ever shall* be known of the nervous system, in *every possible condition*, which is rather a bold position to assume, and what few who have studied the subject will be disposed to accord even to the gifted individuals referred to. Others assume the non-expression of feeling was a mere matter of stoicism, and the general inference to be deduced from the whole harangues of these parties is, that the whole was a piece of collusion and deception. Had the parties intended collusion and deception, would they have admitted that the patient heard the sawing of the bone, or groaned or moaned during the operation? One gentleman, the learned editor of a medical journal, I think, admitted he was bound to believe the testimony of those who had brought the case forward, but frankly avowed, that for his own part, " *he would not have believed it, although he had seen it himself.*" When a man has attained to this state of prejudice and incredulity, of course it would be idle to adduce to him either experiment or argument.

I would beg respectfully to ask, Had the mind of any of these gentlemen never entertained the possibility of a patient, long accustomed to severe pain, moaning from habit, whilst free from pain at the moment; or even feeling pain, and manifesting the same by sensible signs *during sleep*, and yet being quite unconscious of it when he awoke? Do we never meet with similar results in consequence of accidents, in the course of disease, or as the effects of overdoses of narcotics? That such is the case during the artificial sleep induced by the methods I have pointed out in this treatise, I am quite certain. I am equally certain that the sensibility to pricking, and pinching, and maiming the rigid limbs, is gone, some time *before* hearing disappears. Even a piece of paper may be inserted, and retained under the eyelids, without the slightest inconvenience, not even inducing nictitation. In short, I am quite certain that a patient may be sufficiently sensible to hearing to enable him to answer questions, whilst unconscious of pricking, pinching, or strong shocks of galvanism passed through the arms, and that even when roused sufficiently to give expression to feeling such inflictions, if allowed to remain quiet a little afterwards, so as to fall

[1] See Appendix I., Note 13.

cation in the cure of disease, and to theorize at some future period, when we have more ample stores of facts to draw inferences from. From the first I was of opinion, that much of the excitement and many of the phenomena developed, were attributable to the altered state of the circulation in the brain and spinal cord,

into the profound state again, he may have lost all recollection of such in-flictions when roused and fully awake.

From the circumstance of the patient having heard, as it is alleged he did, the sawing of the bone, I am of opinion the operation was commenced sooner than it *should* have been ; and I think it very probable that the moaning referred to might have arisen from a slight feeling of pain, but not sufficient to arouse the patient, or to impress him sufficiently to en-able him to remember it when awake.

In conclusion, from the numerous opportunities I have enjoyed of wit-nessing analogous results, in the course of my operations in Neuro-hypnotism, if I may venture to give an opinion in this matter, I have no hesitation in expressing my thorough conviction that Mr Topham, Mr Ward, and the patient, have all spoken and represented the case with the utmost good faith and candour.

To those who wish to stifle investigation, and hold we ought to rest satisfied with the decision of the French Commission, I beg to remark, that a commission of the same learned body was appointed to investigate and to report on Harvey's discovery of the circulation of the blood, and that this most important discovery was rejected by them as a fallacy. Did their decision alter the laws of nature, or prevent the ultimate triumph of our immortal countryman ? And when so much in error while investigating the more apparent and demonstrable one of the circulation of the blood, is it not quite as likely that they may have been mistaken in their decision on the still more abstruse and subtle subject of the laws and distribution of the nervous influence ?

It is matter of history, in respect to the profession in our own country, that there was not a medical man in England, who had attained forty years of age, who would believe in the truth of Harvey's discovery. Is it to be wondered at, then, that Hypnotism should meet with opposition at the present time ?

To conclude these remarks in respect to this operation : the fact that patients have been known, in some few instances, from natural causes which were not understood, to have undergone severe surgical operations without any sense of pain, instead of militating against the truth of the insensibility of the patient whose limb was amputated during the nervous sleep, tends directly to confirm it ; for if such a remarkable state can exist from some accidental circumstances not understood, there is no reason why a similar condition may not be induced by artificial means.

and especially to the greater determination of blood
to them, and all other parts not compressed by rigid
muscles, arising from the difficulty, during the cataleptiform state, of the blood being propelled in due proportion through the rigid extremities. I have not yet
seen occasion to alter this opinion ; but rather to conclude, that the ganglionic, or organic system of nerves,
is *also* inordinately stimulated from the same cause, and
thus having acquired an undue preponderance induces
many of the remarkable phenomena which have been
referred to. Whoever examines carefully the injected
state of the conjunctival membrane, and of the capillary
circulation in the head, face, and neck, the distended
state of the jugular veins, the hard bounding throb of
the carotid arteries, and the greatly increased frequency
of the pulse, during the rigid condition of the limbs,
cannot fail to perceive that there is great determination
to the head. Again, when all these symptoms are so
speedily changed on reducing the cataleptiform condition of the limbs, how can it be doubted that the rigidity
of the limbs, and consequent obstruction to free circulation through them, is the chief cause of the determination to the head and other parts not directly pressed on
by rigid muscles?[1]

[1] In reference to the cataleptiform condition, I beg leave to offer the
following remarks *merely by way of conjecture*, and with the hope
that they may excite others to direct their attention to the investigation.

Muscular contraction or motion is voluntary or involuntary. The
voluntary arises from a mandate of the mind, proceeding from the brain,
and effecting contraction or shortening of the muscular fibres ; the involuntary, or reflex, from irritation conveyed to the spinal cord, producing
a like result, and may be excited by tickling, pricking, or pinching the
skin of the extremities of a decapitated or pithed animal. It appears to
me, however, that much of the efficiency and tendency to muscular contraction is dependent on another cause, namely, the state of *tone* or *tension*

In conclusion, I beg leave to remark, that the
varieties which are met with as regards suscepti-
of the muscles when considered to be in a state of quiescence ; and this
state of tone I consider depends on the ganglionic or organic system of
nerves. Supposing, from deficiency of this, the muscular system is
relaxed, a morbid tendency to reflex action will be induced, as a musical
string will be more easily excited to vibrate if *moderately* tight, than if
drawn *very tense*. It will also render muscular effort less efficient and
certain, because part of the muscular contraction, which would have been
efficient as available force or motion, will be expended in bringing up the
muscular structure to that state which ought to have been its *normal
condition* of tension or tone.

On the other hand, supposing the organic system has been extremely
active, and rendered the muscular tone *abnormally great*, it will produce
the very reverse effect of that just referred to. It will not only offer
resistance to reflex motion, but also to *voluntary* motion ; and, if carried
to a certain extent, may render the parts fixed and rigid, from the ganglionic system overpowering the cerebro-spinal system.

That this is not mere hypothesis seems to me to be in some degree
proved, by the result of operations referred to in my paper in the
Edinburgh Medical and Surgical Journal for October, 1841, where
muscles which had been rigidly contracted, and had lost all power of
motion, had motion restored by dividing the tendons, and allowing a new
portion to grow between the divided ends, thus elongating the muscles ;
and, in other cases, where there was paralysis *from relaxation*, power was
regained by *cutting out a portion of tendon*, bringing the divided ends
together, and ensuring their adhesion, and thereby shortening the muscles,
and giving them *artificially* that tone or tension, the want of which I considered was the great cause of the continuance of the paralysis. It there-
ore appears to me, that during the hypnotic state there is a complete
inversion of the ordinary condition, and that the force of the ganglionic
system becomes predominant, instead of being, as in the ordinary condition, only subordinate.

Another argument in favour of this view is the well-known fact, that all
voluntary motion, or reflex muscular action, speedily exhausts the powers,
and renders the subject unable to continue such efforts, and fatigued in
consequence of them. Voluntary effort also is *strongest at first*, and
gradually becomes weaker. The functions of the organic system of
nerves, on the contrary, are more equable and persistent in their nature ;
and, although they may be influenced in some degree as to the activity
of their functions, by directing attention in a particular manner,—as the
secretion of saliva by thinking of food, the secretion of milk by the nurse
thinking of her child, etc., etc., still they cannot be said to be under
voluntary control in the same direct manner and degree as muscular
motion. The cataleptiform state induced by Hypnotism comes on gradually. For some time voluntary power predominates ; but at length the

bility to the hypnotic impression, and the mode and degrees of its action, are only analogous to what we experience in respect to the effects of wine, spirits, opium, the nitrous oxide, and many other agents. They are all well known to act differently on different individuals, and even on the same individuals at different times, according to the condition of the system ; but who calls in question the reality of their effects merely because of that want of uniformity of action ?

involuntary rigidity, or organic tonicity gains the ascendancy ; and, although persisted in for a great length of time, is followed by no exhaustion or fatigue.   On the contrary, so far as I have carried the experiments, the whole functions seem to be invigorated by the continuance of this condition.

NEURYPNOLOGY

HAVING in the former part so far explained the mode of inducing the phenomena, I now proceed to detail the cases in which I have successfully applied this process in the cure of disease. I shall endeavour to explain my modes of operating in different affections, so as to enable others to apply with advantage in their practice, what I have found so eminently useful in my own.

When the artificial state of somnolence has been induced in the manner already pointed out, pp. 109, 110, the manipulations must be varied according to the peculiar object we have in view. If the *force* of the circulation in a limb is wished to be diminished, and the *sensibility* also to be *reduced*, call the muscles of that member into activity, leaving the other extremities limber. On the other hand, if the force of the circulation and sensibility are wished to be *increased* in a limb, keep *it* limber, and call the *others* into activity, by elevating and extending them, and the desired result will follow. If *general depression* is wanted, after one or two limbs have been extended for a short time, cautiously reduce them, and leave the whole body limber and quiet. If *general excitement* of the system is wanted, extend the *whole* limbs, causing the patient to call the muscles into strong action, and very speedily they will become rigidly fixed, and the

P   225

force and frequency of the heart's action, and determination to the brain, as evinced by the action of the
carotids, distended jugulars, flushed face, and injected
eyes, will speedily become apparent. By applying the
ear over the region of the heart, it will be apparent that
the force and frequency of the heart's action becomes
prodigiously increased in a very short time after extending the limbs. It will also be found, they may be
very speedily altered and brought down by reducing the
rigidity of the limbs. The difference of rise in the pulse
when extending the limbs *during hypnotism*, from what
happens in the natural state, is one of the strongest
proofs of the patient being in the hypnotic condition.

It has appeared to me, that we have thus the power
of subjecting the brain and spinal cord, and whole
ganglionic system, to a high state of excitement, as
the pulse may speedily be raised to double its natural
velocity, in most cases, and still more speedily reduced
to the natural standard again. Its volume and tension
may also be equally rapidly increased or diminished.
It is therefore naturally to be expected, that the
functions must be greatly influenced by such transitions. Every medical man knows that chronic nervous
disorders of the most painful nature may have resisted
every known remedy for weeks, or months, or years,
but have speedily vanished on the accession of some
acute attack. Now, my views were, in such cases, to
induce an *intense* state of excitement for *a short time*,
to be terminated abruptly, with the hope of changing
the former action, and thus terminating the disorder ;
and assuredly, in many instances the most obstinate
chronic functional disorder is gone, or greatly meliorated,
by a few such operations.

Then, again, by keeping any particular organ awake or active, whilst the others were asleep, I considered there would be a great increase of activity induced, by the whole nervous energy, or sensorial power, being directed to that point ; or by keeping all the other organs active, whilst one which had been *too* active was allowed to remain in the torpid state, that in-ordinate activity would be reduced in intensity, and that probably permanently, — that the inordinate stimulus, in one case, would remove the susceptibility to lower impressions, which were frequently exciting, or habitually keeping up morbid feeling or action ; and in the other cases, that by suspending the morbid sensi-bility of a part for a time, and rousing antagonist func-tions, such condition might be permanently improved.

Whether I have been right or wrong in my theo-retical views, there can be no doubt of the fact that in many instances I have been successful in the applica-tion of Hypnotism as a curative agent ; and the beneficial results of the operations have been so im-mediate and decided, as to leave no doubt that they stand in the relation of cause and effect. However, that much of the success depends on the impression arising from the altered condition of circulation, seems to me to be proved by the fact, that in cases where the sleep was induced *without* the *rise* in *the force and frequency in the heart's action*, by insuring this con-dition, the beneficial result has instantly followed, where there has been no previous improvement with the *low* pulse. The following is a remarkable instance of this :—Nodan, deaf mute, twenty-four years of age, was considered never to have heard sound excepting the report of a gun or thunder, when there was succussion of the air sufficient to induce *feeling* rather than hearing, properly so called. The mother told me Mr Vaughan, head master of the Deaf and Dumb Institution when Nodan was at school, considered any indication of hearing referred to was *feeling*, and not hearing, properly so called. At the first operation there was very little rise of pulse, and afterwards I could not discern he had any sense of hearing whatever. At next trial the pulse was excited, and so remarkable was the effect, that in going home he was so much annoyed with the noise of the carts and carriages, that he would not allow himself to be operated on again for some time. He has only been operated on a few times, and the result is, that although he lives in a back street, he can now hear a band of music coming along the *front* street, and will run out to meet it.

I shall first illustrate the efficacy of hypnotism on the various senses, and also on the mental condition. And first, of sight. The mode of operating in chronic cases, is first to induce the sleep, then extend the extremities, and keep the eyes from getting into the torpid state, by fanning them, or passing a current of air over them occasionally. The length of time required to keep such patient in this condition may vary from six to twelve minutes, according to the state of the circulation. The following cases will illustrate the affections of the eyes in which I have applied this mode of treatment with advantage.

Case I. Mrs Roiley applied to me on the 6th April, 1842. She stated she was 54 years of age ; that for the last sixteen years she had been a great sufferer from an affection of the head, attended with pain in the eyes and weakness of sight ; that it was now become so bad, that she could not continue to read for more than a few minutes at a time, even with the aid of glasses. She had undergone the most active treatment under first-rate medical men, including bleeding general and local, blistering—on one occasion, she was twice bled with leeches, and had five blisters to her head in one month—and almost every variety of internal medicine which could be suggested for such a case ; but still without improving her sight. For years she had required to have her head shaved every few weeks, and cold effusions and spirituous lotions frequently applied to it, to reduce the excessive heat and other uncomfortable feelings. The skin of the palms of the hands was so hard, dry, and irritable, as to render it liable to chap whenever she attempted to open the hands fully. The pain during the day, and general irritability, had rendered it necessary for her to take a composing pill three times in twenty-four hours, for some time ; still her rest was so bad as to force her to rise and walk about the room several times during the night ; and her memory had become so much impaired, that she often required to go upstairs and then down again several times before she could remember what she went up for. About three years before consulting me, she had a paralytic attack, which deprived her of power of the muscles of the right side of the face for a few days. Such had been the general state of her health before consulting me, and the state of her *sight*, and the result of my operation will be understood by the following document, which is attested by herself and others, who were present when I first operated on her :—

" Mrs Roiley (aged 54), Chapel Street, Salford,

formerly of South Windsor Street, Toxteth Park, Liverpool, as Miss Robinson (four years ago), has been gradually losing sight since thirty-eight years of age. Called on me for the first time, 6th April, 1842. Could not read the heading of the newspaper, excepting the words, ' Macclesfield Courier '; after being hypnotized for eight minutes, she could distinctly read 'and Herald,' and in a few minutes more the whole of the smaller line, ' Congleton Gazette, Stockport Express, and Cheshire Advertiser,' also the day, month, and date of the paper. That the above is a correct report, is attested by the patient herself and other three patients, who were present the whole time.

(Signed)   ALICE ROILEY.
                   M. A. STOWE.
                   ANN STOWE.
                   HENRY GAGGS."

When Mrs Roiley called on me two days after, she gave me the following report. After leaving my house on the 6th, she was much gratified to find her sight so much improved, which induced her to go and test it by looking at articles displayed in shop windows, and in particular remarked that she had walked up to Mr Agnew's shop window, and was able to see distinctly the features of a portrait of Sir Robert Peel, and to read under it, " Sir Robert Peel, Bart.," *without* her glasses, neither of which she could have done for long before. She also stated, that after being at home, she took up the small diamond Polyglot Bible, and with the aid of her glasses, was agreeably surprised to find she was enabled to read the 118th

Psalm (29 verses), although this had been, as she expressed it, a sealed book to her for years. The following is the report which was recorded and attested by her on the 12th April, 1842 :—" Mrs Roiley was able to read a Psalm with the aid of her glasses in the smallest sized Polyglot Bible same afternoon she was first hypnotized. Two days after (8th April), was hypnotized a second time. Next day, made a net handkerchief with the aid of her glasses. April 12th, has gone on improving, and in my own presence and several others, with the aid of her glasses, read the Polyglot Bible with ease and correctness, which she said, had been a sealed book to her for years before I operated on her.

(Signed)   ALICE ROILEY.
M. A. STOWE.
WM. HALLIDAY."

It is gratifying to be able to add, that the improvement of the sight has been permanent ; and not only so, but that the whole painful catalogue of complaints with which she had been afflicted speedily disappeared, namely, pain of the chest, head, and eyes, loss of memory, disturbed sleep, irregularity of the secreting and digestive functions, and instead of the arid skin, regular action of it, so that the palms of the hands, which were so harsh and arid that she could not extend them without lacerating the skin, causing great pain and annoyance, were very soon as soft as a piece of chamois leather.[1] The whole of this improvement

---

[1] Very lately, a lady about 25 years of age was hypnotized by me. On being roused, she expressed her surprise to find her hands bathed in perspiration, as she observed *she was never known to have the slightest moisture on her hands till that moment.*

was accomplished entirely by this agency, as she had no medicine whatever during her attendance on me; nor has she required any up to this date, 20th February, 1843, when I read this report to her, and when she remarked it was much *under* drawn; that with great truth I might have represented her as having been a greater sufferer.

Mrs Roiley is a very intelligent person, and one whose Christian profession and principles place her statements above all suspicion. She has been seen by many eminent professional and scientific gentlemen, who can bear testimony that they have had from her own lips the same statements as I have recorded above.

It appears to me that it would be impossible to adduce a more striking proof than this case affords, of the great and undoubted benefit resulting from the application of any remedial measure. The improvement was so remarkable, as to admit of no doubt as to its reality, and so immediate after the hypnotizing, as to prove they stood in the relation of cause and effect, no other remedy being in operation; and whatever may be supposed capable of being achieved through the mere power of imagination, as regards *certain* functions, the sense of *sight* could scarcely be supposed capable of being so much meliorated directly through that influence.

Case II. is that of Mrs M. A. Stowe. This lady was present when I first operated on Mrs Roiley, and was so much gratified by the effects she witnessed in that case, as to induce her to consult me as to the state of her own eyes, and the probability of benefiting them by a similar operation. Mrs Stowe was 44 years of age, and had experienced such weakness of sight as to require the aid of glasses for the last twenty-two years, to enable her to sew, read, or write, and, for some years past, she required them to enable her to transact her most ordinary household duties. The following is the statement of her condition, which I noted at the time, and is attested by her own signature, and that of others then present :—

"Mrs Stowe, aged 44, 1 Bank Place, Red Bank, Manchester, has been troubled with weakness of sight for twenty-two years, so as to require glasses to enable her to read or sew. When tested to-day, 8th April, 1842, without her glasses, could not distinguish the large (capital) letters of advertisements in a newspaper, nor large heading of the paper. After being hypnotized for eight minutes, she could distinctly read both the large and small heading, and day, month, and date of the paper.   (Signed)   M. A. STOWE."

"She has also been able to sign her name to attest the accuracy of the above statement, before her daughter, and another patient.   (Signed)   ANN STOWE."

"10th, Called on me, and informed me she had been able to make herself a blonde cap, and to thread her needle *without* spectacles,[1] which she could not do before for twenty-two years. 12th, Continued improving; told me she had been able to write up her accounts *without* glasses.   (Signed)   STOWE.
                                      WM. HALLIDAY.
                                      ALICE ROILEY.
                                      ANN STOWE."

[1] I have myself seen her thread a No. 8 needle on several occasions.

This patient has retained the improvement of her sight. She has also informed me, that she was agreeably surprised, after she left my house, the *first day she was operated on*, to find, as she went along the streets, that she could read the *sign-boards*, which she could not do for years before. She has also named to many others, as well as myself, a very convincing proof of her great improvement in this respect. Before being operated on by me, on the 8th April 1842, if she went a-shopping, *without her glasses*, she was sure to make some mistake as to the quality of goods purchased, and have the trouble of going back to have them exchanged, but now she never requires to take her glasses with her, as can be testified by the shopmen where she makes her purchases. Her memory and general health have also been greatly improved by the same operations.

Case III. Miss Stowe, daughter of the former patient, 22 years of age, "was under the necessity of reading, and doing any particular work, with the aid of glasses, for the last two years, but has never required them since she was first hypnotized, and can now read the small Polyglot Bible." This is attested by her mother, herself, and Mr William Halliday, and Mrs Roiley.

The improvement has been permanent, and she has threaded a No. 12 needle in my presence, eight months after I first operated on her.

Case IV. Mr J. A. Walker, 22 years of age, had always had very weak sight, but since being hypnotized has been greatly improved in his sight, as well as in his memory and general health.

Case V. Mrs C., aged 83, had, from her age, required the use of glasses for many years, to enable her to sew or read. Last August I hypnotized her for deafness, with very decided advantage, and I told her I also expected to improve her sight at the same time. She was very incredulous, but was agreeably surprised to find, that after a *second* operation she was not only able to *hear* much better, but also to sew some flannel, threading her needle *without* her glasses. She had been thus occupied for several hours, when I called to see her, after the *second* operation.

There have been cases in which I have tried this method without success, but this proves only that we must never expect to obtain possession of a universal remedy. Cases of confirmed amaurosis, which had resisted every other known remedy, and which were only undertaken by me at the desire of the patients, and sometimes of medical men also, as a forlorn hope, have, as in most cases was suspected might be the result, proved unsuccessful, and, through these, attempts have been most ungenerously and unwarrantably made to throw discredit on the power of hypnotism altogether. It has proved successful in too many instances, however, to be borne down by such paltry and pitiful misrepresentation. I could easily adduce many more successful cases, did I deem it necessary, but shall only give two more.

Case V. Mr J—— has always had imperfect vision, is near-sighted, has strabismus of right eye, and the sight so dull, that it was with great difficulty he could, without glasses, see the large letters (on white paper) in the title page of the " Medical Gazette." After the first operation he could see better, and after it had been repeated a few times he could, without glasses, read a few words of the leading article of that work, and after a few more operations, could read the type in

which the lectures, at the beginning of the work, are
printed.

Case VI. Mrs S., one of my own near relatives, had
a severe rheumatic fever in January, 1839. During
the course of this disease the left eye became implicated, involving both the internal and external structures of the organ. She had the benefit of the advice
of one of the first-rate oculists in Edinburgh. She
was under his care till the August following, when
he considered farther attendance unnecessary, but gave
such instructions as he deemed expedient for her future
management of it, and which had been duly attended
to till the period when I first saw her, in June 1842.
At that time she came on a visit to my house. The
eye was free from pain, but was of no service as an
organ of vision. There was an opacity over more
than one half of the cornea, sufficient to prevent distinct perception of any object placed opposite the
temporal half of the eye, all being seen through a
dense haze ; and objects placed towards the opposite
side were seen very imperfectly, owing to the injury
the choroid and retina had sustained in the points on
which the images of such objects were reflected. The
opacity of the cornea was not only an obstacle to
distinct vision, but was also a source of annoyance,
from its disfigurement, being obvious even to those at
a considerable distance.

Notwithstanding the great advantage I had seen
other patients, afflicted with affection of the eyes,
derive from hypnotism, it never occurred to me that
such a case as that of Mrs S. was likely to be benefited
by such an operation. I had, however, recommended
it to her for a severe rheumatic affection of the right

shoulder and arm. She had been in my house about three months before she could make up her mind to undergo the operation, but at length, the violence of the pain impelled her to try it, or anything else I should recommend. I of course hypnotized her, which immediately relieved her pain so much, that after the first operation, she could move the arm freely. The operation was repeated the following day, with complete relief as regarded the arm; and to the surprise and delight of the patient, myself, and others present, she found her *sight* so much improved as to be able to see every thing in the room, and to name different flowers, and distinguish their colours, whilst the right eye was shut, which she had not been able to do for more than three years and a half previously. I consequently now repeated the operation daily, and, in a very short time, had the satisfaction of seeing the cornea so transparent, that it requires close inspection to observe where any opacity remains. Neither external nor internal means were used during this improvement, nothing but the hypnotizing was had recourse to; and during the three months I had an opportunity of watching it prior to these operations, there was no visible change in the condition of the organ. I should observe, that after the first operation, there was considerable smarting in the eye, which continued all night, and, in a less degree, after future operations, which, no doubt, roused the absorbents, and effected the removal of the opacity of the cornea. Stimulating the optic nerve to greater activity, however, must have been the chief cause of the very rapid improvement, which enabled her to see objects after second operation. I should remark, that the sight, with regard to objects seen from the *temporal* side of the eye, is much more distinct than from the nasal side, owing to the retina and choroid having sustained irreparable damage during the inflammatory stage at the commencement of the attack in 1839.

Case VII. Mr Holditch, 39 years of age, had been partially paralytic for ten years, which came on some time after a fall. Shortly after the fall, he experienced an attack of double vision, which went off after bleeding, blistering, and the usual treatment, but was followed by paralysis of the lower limbs, which induced him to consult me on the 18th of February, 1843. See Case XXVII., p. 278. He was very much surprised, when I told him he had defective vision of the right eye, said he was not aware of it, *and would not believe that I was not mistaken, till I tested him*, when he found he could barely see the capitals of the words, " Medical Gazette," as heading of the leading article of that work, whilst he could read the ordinary size print of the page with the other eye. After being hypnotized, I tested him in the same position, and with the same degree of light, and he could then read the *small sized print with it*, and it has continued so ever since. He could also walk across the room without crutch or stick, which he could not do before, at which he was very much surprised, as he was quite conscious the whole time, and therefore could not believe any good could have resulted to him from what was done, till he had the positive evidence of it in being able to see and walk.

Here, then, we have seen three cases of improved vision consequent on hypnotizing for other affections, and where, consequently, the improvement could not at all be attributable to imagination, but to the altered condition in the capillary circulation and distribution of the *vis nervosa.*

In cases of active inflammation of the eyes, either external or internal, I have never tried hypnotism. By the mode calculated to excite the circulation, of course it would be quite inadmissible; and it could only be speculation for me to hazard an opinion as to its probable result by the other mode.

The extraordinary excitement of the auditory organ, which I had observed in the course of my early experiments, and the fact that hearing was the last sense to disappear during this artificial sleep (unless we except that of the sensibility to a current of air), led me to anticipate most satisfactory results from this process in the treatment of deafness, arising from torpor of the auditory nerves. I consequently tried it in such cases, and where there has not been destruction, or irreparable organic injury to the auditory apparatus, I can confidently say, I know of no means equal to hypnotism, for benefiting such cases. Of course, it cannot suit *all* cases, but I am satisfied it will succeed in a numerous class of cases, and in some which bid defiance to all other known modes of treatment.

I am enabled to state this confidently, not only from my own personal success, but also from that of others who have fairly tried it. One professional friend, Mr Gardom, introduced to me two patients whom he had improved so much by hypnotism only, that they were enabled to hear the sermons of their respective pastors, which they could not do before, in consequence of which one of them had to leave her favourite minister, and go to another church; but, after being hypnotized, has been able to hear so much better, that she has been thus induced to return to her *former* pastor.

The great success which I had experienced from hypnotism, in improving those who were deaf through disease, led me to hope it might be of service to some of those who were born deaf and dumb, and I therefore tried it in such cases with a considerable degree of success, ultimately with a success beyond my most sanguine expectations. In consequence of what had been done and exhibited at my lectures, the medical profession of Liverpool, to their credit be it recorded, recommended to the governors of the Deaf and Dumb Institution there, to permit an experimental trial to be made at their Institution. The governors refused their assent to this *within the walls of the Institution*, but agreed to permit a trial to be made with such out-door pupils as could be induced to submit to it elsewhere, the consent of the parents having been obtained. In consequence of this, a committee of the governors and the medical faculty was appointed to superintend the said investigation, and I was invited to go over and conduct the experiments in their presence, and it was proposed a report of the results should be published in the Medical Journals, at the termination of our labours. The difficulty of getting the pupils and their parents to attend, induced us to abandon the proceedings after two trials had been made, so that it would be quite inconsistent with the conditions stipulated, at the commencement of said investigation, to publish any report of the result of this *partial* investigation. However, I think I cannot better illustrate the extent of my expectations, in reference to such cases, than by transcribing an extract from my address to the said committee, prior to commencing our experimental trial.

" Hitherto, these patients have been considered beyond the pale of human aid, so decidedly have they resisted all means tried for their relief; and the morbid condition of the organs, as ascertained by dissection, was sufficient to warrant the inference that it was *improbable* any remedy could ever be discovered for such cases. Fully aware of this pathological difficulty, I was nevertheless inclined to try the effect of neuro-hypnotism with congenital deaf mutes, knowing it could be done with perfect safety, and without pain or inconvenience to the patients. Moreover, from having witnessed its extraordinary power of rousing the excitability of the auditory nerves, I entertained the hope that it might thus be capable of exciting *some* degree of hearing, from the increased sensibility of the nerves compensating for the imperfection of the organ. I was not, and am not even now, so visionary, as to expect *perfection of function*, when there is great *im*perfection of the organ. Perfection of organization and function must be co-existent; at least the function cannot be *perfectly* performed when the organization is *much* impaired. The result of my first trial was beyond my most sanguine expectations, which induced me to persevere, and the result has been, that I have scarcely met with a case of congenital deaf mute, where I have not succeeded in making the patient hear in some degree. Many may never hear so well as to make it available to holding conversation by its aid; but still it is most interesting in a physiological point of view, to know the fact, that by this means the imperfect organ can be roused to *any* degree of sensibility to sound, as even this must tend to the improvement of the general functions of the brain, rather than being *entirely* deprived of one source of its appropriate stimuli.   I have no doubt, moreover, that many cases will, by this means, be restored to such degree of hearing as will be available for colloquial intercourse in society, which never could have been accomplished by any other means hitherto tried.   If my success with the cases assembled here is at all equal to what it has been with others elsewhere, I think it cannot be otherwise than gratifying to you to find that our art has acquired a new and important power in this agency.   I must not, however, omit to add, that many cases may show no improvement at a *first* or *second trial,* and yet be very satisfactory after a few trials.   According to my experience, there is much greater chance of benefiting *congenital* deaf mutes, than those who have become so from disease or accident, to *the extent* of *total loss of hearing.*

"In testing patients as to their power of hearing, I consider it quite necessary to adopt a different plan for those who are *congenital* deaf mutes, from what we do with those who have known what perfect hearing was at some former period of their lives.   It is quite true that the latter class may be unable to hear a musical box, or the tick of a watch, when held at a little distance from the ears, but can hear it when pressed *against* the ear, or the mastoid process, or against the teeth, owing to the greater conducting power of the bony structure.   There are patients of this class, however, who declare they have no sense of sound when so tested, because their previous knowledge of the sense enables them to distinguish betwixt *hearing, properly* so called, and *common feeling.*   In testing *congenital* deaf mutes, from their want of this previous knowledge, they will all signify they hear, if any sonorous or vibrating body is pressed against the ear. This, however, I do not consider we have any proof of being *hearing*, but *feeling;* because they had no previous knowledge to direct them as to the peculiar sensation of *correct hearing;* and they will give the same indication if the sonorous body is placed on any other solid part of the body, according to its respective degree of sensibility. In applying tests to *congenital* deaf mutes, therefore, I consider they have no sense of hearing, if they cannot hear the sound of a musical box *held close to* but *not touching* the ears, or any other sonorous body whose vibrations do not excite such oscillation in the air as is sufficient to be recognized by *common feeling*. It ought also to be borne in mind that the *common* feeling of the deaf and blind is generally much more acute than in those who have not been deprived of those senses. At all events we cannot err in taking this as our standard, because, if those who did not hear on the application of such a test *before* the operation, do not hear it also *after* the operation, we shall consider there is no improvement; and if those who hear it at a certain distance *before* the operation, cannot *after* the operation hear it at a *greater* distance, it must also be considered no improvement has been made. But if the former can, *after* the operation, hear *without the box touching* the ear, and the latter can hear *at a greater* distance, then of course we are entitled to say an improvement has resulted from the operation."

These extracts should be sufficient to explain what the extent of my expectations were as to meliorating the condition of *congenital* deaf and dumb patients, the principles upon which these expectations were based, and my mode of testing the original and subsequent condition of such patients. The following cases will prove that my anticipations have been so far realized— in one case to an extent I never calculated on. The mode of operating is, hypnotize the patient, extend the limbs, and gently fan the ears.

Case VIII. The case of Nodan has already been referred to at page 227, and I shall therefore merely add here, that he was 24 years old, was never considered to have had the power of hearing, properly so called, according to the opinion of the head master of the Deaf and Dumb Institution, where he was a pupil; that *after* the *first* operation I satisfied myself *he had no sense of hearing*, but after the second, which I carried still farther, he *could* hear, and was so annoyed by the noise of the carts and carriages when going home, after that operation, that he could not be induced to call on me again for some time. He has been operated on only a few times, and has been so much improved, that although he lives in a back street, he can now hear a band of music coming along the front street, and will go out to meet it. I lately tested him, and found he could hear in his room on the second floor a gentle knock on the bottom stair. His improvement, there-fore, has been both decided and permanent, and is entirely attributable to hypnotism, as no other means were adopted in his case.

Case IX. "Mr John Wright, Pendleton, 19 years of age. Congenital deaf mute. Was four years at the asylum under Mr Vaughan. Never heard sound. On testing, could not discern the tick of a watch pressed against the ears, nor a musical box, *unless when pressed against the ears*, which was evidently *feeling*, *and not hearing*, as he evinced the same expressions when it was applied to the shoulder, chest, or back of the hand. After being hypnotized for eight minutes, he could hear the musical box held *more than an inch from* the *left* ear, but not at all with the *right*, if not pressed against it, which was of course only feeling. Certified as correct by the father of the patient.

(Signed)   JOHN WRIGHT."

" MANCHESTER, *8th April*, 1842."

" After writing the above statement, he was again tested, and could hear the box *half an inch from the right* ear.   (Signed)   JOHN WRIGHT."

The latter fact, of hearing better after being roused than at the very moment they are roused, occurs in cases generally. This patient attended daily for a short time, and made considerable progress in the power of hearing, but like too many others he had not patience to persevere, which his father, who is a very respectable and intelligent man, wished him to do. Unfortunately the deaf and dumb are not aware of the *extent of their privation*, or of the real advantage they would obtain by persevering, and their expectation, and that of their friends, in most cases seems to be, that the moment they have the power of *hearing* restored in some degree, they should, as by a miracle, also be immediately inspired with the gift of tongues, and be able to speak and understand language without study, toil, or trouble. This has been so well expressed by John Harrison Curtis, Esq., that I shall quote a paragraph from his pen on the subject.

"Kramer condemns the cases recorded as cures by Itard, Deleau, and others, because, when published, the patients had not acquired a facility of speech equal to that evinced by other people of the same age; forgetting, that when the deafness has been cured, the individual is placed precisely in the position of a child that has to acquire the faculty of speech, and not unfrequently the power of thought ; while, at the same time, if he have approached the age of puberty, he has to contend with false impressions created by the erroneous perceptions which affected him while unable, from his infirmity, to impart his feelings and ideas to his fellow-creatures ; in fact, he is placed in the same position in regard to hearing as Cheselden's patient was with respect to vision. The organ, when the cophosis is removed, requires to be carefully educated to perceive, understand, and distinguish the variety of sounds which will impinge upon the auditory nerve, a task requiring much time for its accomplishment. The cure of congenital deafness, consequently, may be effected, and yet rendered efféte, for want of this necessary subsequent education."

After remarking that many cases of deaf dumbness arise from disease, and are only partially deaf, he added, " Many of these cases admit of amelioration, some of cure ; and I hold, that wherever there is a chance only of doing good, it ought not to be neglected ; it may certainly raise hopes which may be nullified hereafter, but not in the patient, who cannot comprehend the motives of the proceeding ; nor would the friends be much annoyed thereat, if the surgeon has performed his duty properly, by showing, that although there is a chance of success, it is after all only a chance."—" It does not occasion a loss of valuable time, worthy to be put in competition with the prospect of restoring even one individual to the enjoyment of the society and converse of his fellows."—" Many would be rendered (by proper treatment) useful members of society, who, under the present system, remain hopeless objects of commiseration as long as they live." Mr Curtis farther adds, " I perfectly agree with Dr Williams, who says, a cure ought always to be attempted, and that at the earliest moment at which deafness is detected ; and children so affected should mix with others not deaf, and no symbolical education should take place until all chances of cure are gone." *Medical Gazette*, 23rd September, 1842.

These remarks are so judicious and important as to require no comment by way of enforcing them on any intelligent and candid reader.

The following case having been the cause of much controversy, I shall give it in detail. Before operating on the boy, in the presence of the gentleman who brought him to me, I asked the lad, *in writing*, if he ever heard, to which he returned answer (also in writing), " No." I then proceeded to operate on him, and the following is a report of his case from my note-book.

Case X. " James Shelmerdine, Mr Barker's, 83 High Street, Manchester, aged fourteen years and a half, was born deaf and dumb, and educated at the Manchester Deaf and Dumb Asylum, and came out last June, in consequence of his age. 4th January, 1842, I subjected him to the mesmeric influence, by causing him look at my glass rod, and in thirteen minutes aroused him by a clap of the hands, when he could hear the tick of my

watch applied to the right ear, but only very slightly so
when applied to the left. Could hear me speak loudly,
but could not tell what I said to him. This took place
in presence of his master, who brought him to me, and
now attests the correctness of the above. The boy has
other two brothers deaf and dumb.

(Signed)   MATTHEW BARKER." [1]

"5th January. Again subjected him to the operation.
In twelve minutes he could hear my watch at nine
inches from right ear, and at six from left.

"7th January. Called upon me, and could hear with
the right ear at four and a half inches, and one inch
from left ear. After being hypnotized for ten minutes,
he could hear the watch at seven inches from right, and
at four inches from left ear.

"17th January. After operation could hear six and a
half inches with *left*, and seven and a half with *right*.
20th. Could, after being roused, hear my watch at
seven and a half inches from *left* ear, and at nine inches
from right."

The boy was now tested by competent judges, and
pronounced capable of imitating articulate sounds *without seeing the motion of the lips*. To render this the
more certain, he was tried with a word requiring *no*
motion of the lips and spoken near his ear, which he
distinctly imitated.

I now commenced to teach him to speak a few simple
words, and he got on very well; and that he could do
so very satisfactorily, I considered there was ample
proof by what he accomplished at my lectures. There

---

[1] Mr Barker was not the *boy's* master, but employed some of his friends,
as was afterwards explained to me.

were some who could not believe he could have been born entirely deaf and dumb, when they heard how well he imitated articulate sounds when the motions of the lips were concealed. This was particularly and warmly disputed at a lecture I gave at Liverpool, on the 1st of April, 1842. The boy was asked, without my knowledge, by Mr Rhind, head master of the Deaf and Dumb Institution of Liverpool, if he ever heard before being operated on by me, to which he answered, " No." Next day, in the presence of several friends, I again questioned him in writing as to his original condition, when he gave the following answers, which he certified by his signature as being correct. Fortunately, this document, by the merest accident (having been written on the back of a letter belonging to another gentleman), has been preserved, and I shall here transcribe it *verbatim*. " ' Could you ever hear before I operated on you ? '—' No.' ' How did the master of the school teach you to say, papa, mamma ? '—' Few days.' ' *How* did he do it ? '—' Ba, be, bi, bo, bu.' ' Did the master ask you to watch the motions of his lips ? '— ' Yes.' ' Did he try to teach you to speak by applying his mouth to your ear ? '—' No.' ' Did you ever say what you did to me before ? '—' No.' ' Did you ever read it, so far as you remember ? '—' No.'

(Signed)   JAMES SHELMERDINE."

Hitherto the boy had only been taught single words. The last two questions refer to part of the " Lord's Prayer," in English, which I had been teaching him to speak by *means of hearing ;* and although he speedily made a good attempt at repeating part of it, the effect was so different from that of the mode adopted at school, or that conveyed to his mind through the organ of sight, when reading it, as he must have been accustomed to do, *that he did not know what it was I had been teaching him to speak.* Could a stronger proof than this be adduced that the boy did not learn to speak by *hearing* before he was under my treatment?

I also, on the same day, taught this boy to repeat part of the Lord's Prayer in Latin, to do away with all ground of cavil as to what he *might* have learned at the Institution ; and at my next lecture at Liverpool, the week after, he was heard to be able to repeat it when spoken to him in a moderate tone of voice, whilst the motions of the lips were concealed, and that taking the words in *any* order, so that there could be no ground of mistake as to his *hearing* what he repeated.

Various surmises having now got out, that this boy, James Shelmerdine, *might* have had, or *must* have had, the sense of hearing originally, and that his present condition could not possibly be the result of hypnotism, I addressed a letter to Mr Bingham, who was head master of the Asylum during the five years this boy was at school, requesting him to favour me with information as to James Shelmerdine's *real* condition up to the time when he left school. The following is his reply, and I may add, I am not personally acquainted with Mr Bingham. After describing the partial hearing of this boy, which varied greatly, Mr Bingham adds,—" I never considered his hearing sufficient to distinguish one sound from another in conversation, and consequently, never attempted to teach him to speak in any other way than that which I use with all children born deaf. If hypnotism, or mesmerism, has enabled him to imitate the sounds you wished to communicate to him, without his observing the lips, I do not hesitate to say that you have achieved that which I never could have expected ; and, under such circumstances, I think every encouragement ought to be given to your plan. You would greatly oblige me by saying if this has been accomplished, as *the boy was quite incapable of distinguishing one word from another when he left me, if spoken behind his back.*"

Fortunately I had no difficulty in satisfactorily substantiating this, for, besides having been so repeatedly proved in the public lecture-room, here and elsewhere, he had also been tested before a number of the most distinguished members of the British Association, last June, and, more recently, before a dozen witnesses, including the present head master of the Deaf and Dumb Institution of this town. I instituted this investigation in consequence of some gross attempts which had been made to misrepresent my conduct in reference to this case. The following is an extract from the report of his condition on the 25th July last (1842), and is attested by Mr A. Patterson, head master of our Deaf and Dumb School, and twelve more witnesses :—
" James Shelmerdine was examined at Mr Braid's before the undersigned, in reference to his hearing, and he readily repeated part of the Lord's Prayer, both in English and Latin, both backwards and forwards, after Mr Braid repeating the words in a moderate tone of voice, without being able to see the movement of the lips."

I had not seen the boy for about a month before this investigation, and I would ask, did he not here manifest a decided improvement from the state he was in when he left school, when, as borne testimony to by Mr

Bingham, "he was quite incapable of distinguishing one word from another," if spoken so that he could not see the motion of the lips? and I am quite certain this was his condition *immediately after my first operation.* As has been already stated, he could not then distinguish one word from another, however loudly spoken close to his ear.

After communicating these statements of what the boy *could* do, as recorded at the investigation on the 25th July, Mr Bingham favoured me with a second letter, from which I make the following extract :—
" James Shelmerdine's performance in repeating the Lord's Prayer, in Latin and English, when the motions of the lips were concealed from him, is a convincing proof that he must have benefited greatly by it (hyp-notism), as he could not distinguish one sound from another by oral communication."

The following fact also proves the great improvement in the boy's hearing. One afternoon he was in my hall, when a lady was playing the piano, and singing, in a room up stairs. He seemed so much pleased with the music that I gave him permission to go and hear it. He instantly went up stairs, and into the drawing-room, by himself, and seemed quite delighted with the sound of the music, as several who saw him can testify. This, I am quite certain, he could not have done for some time after he came under my care.

In fine, I feel confident, that had this boy persevered with the operation, and been taken pains with by his parents, to teach him to speak, and understand the meaning of what he spoke, he would, long ere now, have been able to hold oral communication with others with less trouble, and in a more moderate tone of voice than we must resort to with many whom we meet with, who have become hard of hearing from age or disease. It is, however, so much more trouble, at first, for the friends to teach them language, than to hold intercourse with them by signs, that they will not bestow it, and the patients, from not knowing the extent of their privation, can be less expected to exert themselves for acquiring the good they know not ; and therefore, I feel assured there will never be much achieved for the *poor* in this way, unless within the walls of some public institution ; but, that there are many who might be permanently benefited in such situations I have no doubt. In the paper by Mr Curtis, to which I have already referred, he writes thus in reference to the pathological condition of the organ in those born deaf and dumb :—" I am of the same opinion as Itard in this respect, that structural disease does not occasion more than one case in five, leaving, consequently, many cases in which medical assistance may prove of service; and I do not acknowledge that the 'weakness of the nerve, approaching to paralysis, or an actual paralysis of the nerve,' which Dr Kramer assumes to exist in those cases where congenital cophosis is present, and no structural derangement, must necessarily be as in-curable as structural deficiency. We are not apt to abandon incipient palsy of a nerve of sense or motion, in other parts of the system, without an attempt at relief; and I see no reason why the unfortunate being afflicted with deaf dumbness, should be surrendered to his fate, without a well directed attempt being pre-viously made to redeem him therefrom."

This, together with the statement of his experience, ought to encourage farther trials, and especially now

that we have got a new and more powerful agent to
operate with than any hitherto brought into operation
in such cases.   The results of the following case have
far more than realised my most sanguine expectations.
It clearly proves, that persons with perfect organization
may have been deaf and dumb from birth, and continue
so merely for want of a sufficient stimulus to set the
machinery in motion.

In consequence of the remarkable improvement of
hearing, through hypnotism, evinced in the case of
Mrs C. (Case IV. already recorded), I was asked to
give my opinion as to the probability of a similar
operation benefiting a girl who had been deaf and
dumb from birth, and who was sister to a servant in
the family I was then visiting.   I told them what my
experience had been in respect to such cases, and it
was accordingly arranged that I should see the patient,
and try what could be done for her, the following day.

Case X^A. 9th August, 1843.   The girl, Sarah Taylor,
was nine and a half years of age, very small for her age,
and very stupid looking.   The following is the history
of the case, as stated by father, mother, and elder sister.
She was a seven months' child, remarkably small, the
head large for the size of the body, and soft (" like a
bladder full of water "), and it was long before they
expected to be able to rear the child.   As she grew up
they were much annoyed with her not speaking, and by
her paying no attention to what was said to her.   At
last they found that this was not obstinacy, to which it
had been at first attributed.   They now came to the
painful conviction that she was deaf and dumb.   The
father has assured myself, and many others, that in his
anxiety to obtain proof of her having any degree of

hearing, he has "often stood behind her, and shouted
(as he expressed himself) till he was hoarse again,"
without her evincing any sign of hearing; and that
when she was out of sight they were in continual terror
she would be run over by carts or carriages, as she
could not hear their approach. The testimony of the
mother and sister was to the same effect, that they
never could make her hear, or pay any attention by
calling her, when her back was towards them. In such
position they could only make her observe them by
touching her. They all agree, also, in stating, that she
never could speak so as to be understood, till after
being operated on by me, excepting two or three words,
—father, mother, sister, which she had learned from
watching the motions of their lips. I regret not having
had her tested by a musical box before I operated on
her ; but I am quite certain, that *after* the *first* operation she could not distinguish one word from another ;
and I afterwards had the best possible proof of her
never having heard for any useful purpose, as she was
quite ignorant of the name of *any part of her own body*,
or of any person, place, or thing, as is well known to
many who saw her after I had operated on her. After
the third and fourth operation I could manage to make
her speak a few simple words, and also to make a tolerable attempt at following me when singing the musical
scale.

Ten days after the fourth trial, she was tested and
proved able to do this before fifty or sixty highly
respectable witnesses, including many professional
gentlemen. For months past she has been attending
the Scotch Sessional School, and is making very good
progress in learning, and I have no doubt, will prove

to be a clever girl ; she hears so correctly now, as not only to be able to imitate speaking, but also singing. Mr E. Taylor, Gresham Professor of Music, lately afforded a number of my professional and scientific friends a good proof of this, as he composed an extemporary tune which she and other two patients sang correctly, whilst in the state of neuro-hypnotic sleep. She could have done the same whilst awake, and hundreds have witnessed her speak and sing, both when asleep and when awake.

It is curious, that in some who have a very incorrect musical ear, so that they could not be taught to sing the most simple air correctly when awake, can nevertheless be made to do so, when in this peculiar sleep. This was remarkably exemplified in a young lady, whom I wished to be taught a simple air which she might sing by way of exemplification, at some lectures I was to give at a distance, but it could not be accomplished ; she could not follow in tune more than a note or two together ; but when asleep, she can sing any air correctly which I have tried her with. Still, when awake, she cannot do so. For an example of the same sort during natural somnambulism, see pages 296—298, and 309, of Dr Abercrombie's work on the Intellectual Powers. Of one it is noted, " She often sung, both sacred and common pieces, incomparably better, Dr Dyce affirms, than she could do in the waking state." Of the other, "she was, when awake, a dull awkward girl, very dull in receiving any kind of instruction, though much care was bestowed upon her, and, in point of intellect, she was much inferior to the other servants of the family. In particular, she showed no kind of turn for music, and she did not appear to have any recollection of what passed during her sleep." During somnambulism, she sang beautifully, and exhibited great intellectual powers.

I shall conclude this department by recording the following case from my note book. The inability of this patient to sing *in tune* may have been partly owing to a defect in the organ of hearing, and partly to a state of nervousness affecting the vocal organs. The experiment was undertaken merely to gratify the particular desire of the patient, as at that time I had had no similar case, and was not prepared to say, whether it was likely or not to be successful. However, I felt assured it would do him no harm, and made the trial accordingly, and assuredly nothing could have proved more successful or more gratifying than the result.

Case XI. 7th July, 1842, I was consulted by Alexander M'Roberts, 29 years of age, residing with Mr Hannay, of 42 Thomas Street, Manchester. He said, he had never been able to join in tune, although he had frequently attempted to do so. After being hypnotized for some time (about ten minutes), I roused him, and desired him to walk into the dining-room, and after hypnotizing him once more, a friend played the organ, and I directed (or led) him to sing the scale, beginning with D, as he could not sing C, owing to the natural pitch of his voice. He very soon managed to sing the scale quite correctly, upwards and then downwards. I now roused him, and made him sing it when awake, which he did remarkably well. I now tried him with the first part of " Robin Adair," which he followed in correct tune several times. This took place in presence of Mr James Reynolds, Mr Daniels, Mr James Braid, my nephew, and myself. In the evening of that day,

R

after being again hypnotized, he sung the first part of
" Robin Adair" very correctly several times, and also
Pleyel's German Hymn, and the Old Hundredth Psalm,
quite correctly.   Pleyel's German Hymn he never
heard before.   This took place in presence of four
gentlemen.

His inability to sing prior to these operations was
borne testimony to by several of his friends, one of
whom had a good knowledge of music, but despaired
of ever seeing M'Roberts able to sing, and he was
exceedingly surprised at the result.   This patient was
operated on several times afterwards, and when I last
saw him, could sing a considerable number of tunes,
and follow any simple air with ease and correctness.

The next sense I shall refer to is that of smell.
Having put the patient into the hypnotic state, he
ought to be kept in it a longer or shorter time, according
to the object had in view.   If to excite or quicken the
sense, the limbs should be extended and a gentle
current of air should be passed against the nostrils
occasionally ; but if to diminish the sense, this ought
not to be done.

Case XII. is an interesting example of restoration of
the sense of smell by hypnotizing.   A young lady was
subjected to this operation for a different complaint.
On being aroused, and after I left the room, she made
inquiries as to the cause of the great noise she heard
in the house, and expressed her surprise at the noisy
manner in which the various duties of the apartment
where she was were performed.   They assured her
there was nothing going on in the room where she was,
different from what was usually the case, nor was there
any thing to account for the noise she complained of,

and they therefore held her complaints to be only imaginary. She persisted they were real. The fact was, she had been for a length of time dull of hearing, and the improvement of this sense consequent on the hypnotizing, had so quickened the faculty as to account for the difference she experienced. Moreover, she had for a considerable time previously lost the sense of smell, and it was now ascertained *that this sense had also been restored, through the same operation.* Another patient who had lost the sense of smell for nine years, had it restored after being twice hypnotized. For a beautiful illustration of the extent to which this sense is roused during the hypnotic sleep, see footnote,[1]

[1] " A beautifully contrived experiment was here put in practice by Mr Clarke, and Mr Townend, to test the truth of the phenomena. Mr Braid had drawn their attention to the wonderful exaltation of the sense of smell. A rose had been held before the patient, the scent of which she had followed about the platform in every direction with the most excessive eagerness—now standing on tiptoe to reach it when held aloft, anon bending herself forward with the most graceful ease, till her face came almost in contact with the floor—now darting after it across the platform (notwithstanding that her eyes were bandaged) with unerring aim as to the direction in which it was moved—or throwing herself into the most fantastic attitudes, but always with surprising ease, to catch its fragrance when moved merely round her person in tantalizing play. At length she no longer followed it, and Mr Braid now explained that the sense of smell had entirely gone, and could only be renewed by a current of air across the nostrils. Mr Clarke here motioned Mr Townend to go across the platform, which he did very softly, and Mr Clarke then threw the rose to him, a distance probably of from four to five yards. Mr Clarke having thus taken the precaution to guard against the suspicion of collusion or trick, himself passed a current of air across the nostrils of the patient, so as to again exalt the sensibility of the organ. She now moved forward as though in search of some object that had escaped her, and was advancing in front of the stage, which was not exactly in the direction the rose was thrown, when suddenly her limbs and entire body shook with a tremulous motion, and she stooped slightly, and evinced the utmost terror. Mr Braid explained that this was occasioned by the rattling of a carriage over the pavement under the window partly, and partly by a feeling of insecurity, arising from the boards on which she stood being limber, and yielding considerably to the foot. When the noise of the

extracted from a report of my conversazione to the members of the British Association, as recorded by the *Manchester Times.*[1]

The next senses I shall refer to, are touch and resistance ; under which I shall adduce examples of the beneficial results of this agency, in the cure of abnormal exaltation or depression of these functions. There are few diseases more striking in their manifestations, or more important in their character and tendency, than those included in this class, namely, paralysis of sense or motion, or both ; or the reverse, exalted feeling, and tonic or clonic spasm.

Tic douloureux is well known to be one of the most agonizing affections to which the human frame is

carriage had ceased, she turned her face about till it pointed in the direction where Mr Townend stood, when, though he held the rose at a distance of three yards from her, she evidently caught the scent, and darted towards it with unerring precision, and appeared almost to revel with delight in its fragrance. A sudden burst of applause from the audience, quick as thought, dissipated the charm, and she stood aghast, apparently in an agony of terror. Mr —— laughed, and attempted to convey to a small circle around him the impression that all this was feigned, but the attempt was disregarded. In the very front of the company, and amongst those most narrowly watching the experiments, were the Dean of Manchester, the Rev. C. D. Wray, the Rev. N. W. Gibson, the Rev. H. Ethelston, Colonel Wemyss, and a number of others whom we might mention, including several surgeons, who were capable of forming an opinion of their own, and we heard from several of them expressions at once of surprise at the phenomena, and of conviction that they were real. In fact, it was the conviction of common sense, since it would have been far more wonderful as a piece of trickery than as Mr Braid accounts for it. Every one must have felt that it was impossible for any person in a natural state to follow a flower about the stage blindfolded (supposing the patient was awake), passed about as it was from hand to hand backwards and forwards, with such ease, certainty, and rapidity ; but taking Mr Braid's solution of the difficulty, that the senses are unnaturally exalted, the mystery is at an end. The only thing extraordinary that then remains is, that such an agency should not before have been discovered."

[1] See Appendix II., Note 1.

liable. It may arise from a functional disorder of the nervous system, of a local or more general character, or from an organic cause. The symptoms are much the same in both varieties, but the chances of effecting a cure are very different. In the former variety, a cure may be effected, and by no means I know, so speedily and certainly as by hypnotism ; but in the latter, the chances of success are very different, either from this or any other known remedy. I have repeatedly applied it in the one case, without any apparent effect, either good or bad, but, in the other, with the most immediate and striking advantage. I give a few cases in illustration of this success in functional disorder.

Case XIII. W. M'Leod had been suffering for two months from a violent attack of tic of the head and face, which had resisted the treatment prescribed by his surgeon. He had been taking carbonate of iron in ample quantity. After eleven minutes' hypnotism, he was aroused quite free from pain, and it never returned in the same degree of violence, and by a few repetitions of the same process, he was completely cured, and has remained well for about a year. The general state of his health required the aid of other means, but the violence of the tic was overcome before he took a single dose of medicine from me.

Case XIV. A young lady was suffering from a most violent attack of tic douloureux, so much so, that I heard her screams before entering the house. The paroxysms came on so frequently that she was roused before I could succeed in hypnotizing her at first trial. I now administered thirty drops of laudanum, in a little water, sprinkled some over the poultice on her face, and instantly commenced hypnotizing her again. In five

minutes she seemed to be in a comfortable sleep, the
features perfectly placid, the respiration calm, not a
muscle seemed to move during the time I remained in
the room (which was a quarter of an hour), whereas she
had a violent paroxysm every three minutes previously,
contorting her whole body, and when I examined her,
after having been down stairs a considerable time, she
was lying in exactly the same posture as when I left her,
with the same appearance of placid sleep. When I called
next morning I was told she had slept for five hours
and a half, and had had no return of tic after waking.
As she was in the somnolent state, and the paroxysms
of pain suspended *within five minutes*, it is quite clear
this could not be due to the few drops of laudanum,
as they could not have been adequate to arrest such a
violent complaint, at all events, *not in the course of five
minutes.*[1]

[1] The following is the statement of the above case, attested by Mr
Mullard, druggist, who had been called to visit this patient before my
arrival, which I give because of some very unwarrantable interference by
other medical men,—"I was present with Miss G. when Mr Braid
visited her, in consequence of a violent pain in the face, coming on in
severe paroxysms, as occur in tic douloureux. I had applied poultices,
and had other means in readiness, but owing to the violence of the pain,
Mr Braid, the usual medical attendant of the family, was sent for. Her
screams were heard in my house, during the paroxysms, and they recurred
about every minute, and lasted nearly a minute and a half, as nearly as I
can recollect. Mr B. had an opportunity of hearing her on coming into
the house; and shortly after being in her bedroom she had a second
attack. Mr B. now tried to hypnotize her in his usual way, but she was
roused by the violence of the pain. He now gave her a few drops in
water, and sprinkled a few over the poultice, and applied it to the cheek
again, and immediately repeated his operation, after which she seemed to
be in a sound sleep, and gave no farther indication of pain in less than
five minutes. Mr Braid, as well as myself, remained a considerable time,
at least three quarters of an hour, and both left convinced she was comfortably asleep, and next morning I heard she had passed a good night,
having slept about five and a half hours, and that the tic had not returned
since we left. Every word of this has been carefully read and considered
before being signed.   (Signed)   A. T. MULLARD."

21st June, 1842.

Case XV. Miss —— had been suffering severely from tic for several weeks, and had several teeth extracted without relief. During a violent paroxysm, I succeeded in hypnotizing her, and when aroused, it was quite gone, and has never returned.

In the affection to which these cases belong, there is frequently such irritability of the skin, that a slight touch over the affected nerve is quite sufficient to excite a paroxysm of pain. I shall now adduce some cases illustrative of the *opposite* condition, when there was deficiency or entire loss of feeling; and which have nevertheless been greatly benefited, or entirely cured by hypnotism. The following case is illustrative of its successful application where there was paralysis both of sense and motion.

Case XVI. Mrs Slater, 33 years of age, in the autumn of 1841, had suffered a good deal during her pregnancy, and in December of that year was delivered of a seven months' child. From this period, her legs, which had been very weak for some time previously, became very much worse, and in a short time she lost all voluntary power over them, together with loss of natural feeling. She had been under the care of three professional gentlemen, but as she became worse instead of better, notwithstanding the means used, the case had been considered hopeless, and left to itself, for some time previous to my being consulted, which was on the 22nd April, 1842. I found she had not only lost feeling and voluntary motion of her legs and feet, but that the knees were rigidly flexed, the heels drawn up, the toes flexed, and the feet incurvated, and fixed in the position of slight club foot (varus). She had not menstruated since her confinement, but there was no other function

as regarded the secretions or excretions, which appeared
to be at fault. Her speech was imperfect and her
memory impaired. I hypnotized her, and endeavoured,
whilst in that condition, to regulate the morbid action
of the muscles, and malposition of the feet and legs.
In five minutes I roused her, when she thanked God
she *now felt she had feet, could feel the floor with them,
and could move her toes.* I now raised her on her feet,
and with the assistance of her husband supporting her
by the one arm, and myself by the other, she went
across the room and back again to the sofa, moving
her legs and supporting half the weight of her body on
them. I operated on her again the same evening, after
which she was able to support herself standing with the
soles of her feet on the floor. She required merely to
be steadied by placing the points of the fingers of one
of my hands against her back. Before being operated
on, the heels were drawn up, and the feet twisted so
that she could only have touched the floor with a small
portion of the outer edge of the feet, near the root of
the little toes. I hypnotized her in the same manner
daily for some time with increasing improvement, so
that in a week she was able to walk into her shop
alone, merely requiring to steady herself by the wall,
and in two weeks more she could walk into it *without
any assistance whatever.* Two months from my first
seeing her, she went to Liverpool, and was able to walk
several miles in a day. She could walk from the middle
of the town where she lodged, to the pier head and
back, and from her lodgings to Everton and back, all
in the same day, which was several miles partly on very
steep acclivities. She had no relapse, and has continued well ever since.

In a very few days after I first operated on this
patient, the catamenial discharge appeared for the
first time since her confinement. She had no internal
medicine, nor external application whatever to her legs
for several days after I first saw her. Her extraordinary improvement, therefore, resulted entirely from the
effects of the operations. After I had attended her
some days, she required some simple aperient medicine,
and I afterwards prescribed a diuretic, which I hoped
might expedite the cure. The feeling and power of
her legs and feet were greatly restored, her speech
perfect, and her memory much improved, before she
had a single dose of medicine from me. Her improvement therefore was strictly the result of hypnotism only.

The extraordinary effects manifested in this case, as
well as in many others, after a few minutes' operation—
so different from what is realised in the application of
ordinary means—may appear startling to those unacquainted with the powers of hypnotism. On this
account, I have been advised to conceal the facts, as
many may consider it *impossible*, and reject the *less* startling, although *not more true* reports of its beneficial action
in other cases. In recording cases, however, I consider
it my duty to report *facts as I have found them*, and to
make no compromise for the sake of accommodating
them to the preconceived notions or prejudices of anyone.

Case XVII. Samuel Evans, 45 years of age, had
suffered much from pain in the spine, and also been
afflicted with impaired feeling as well as power of the
superior extremities for four years. He suffered also
occasionally in the head, for which he had undergone
every variety of treatment usual in such cases, under
many medical men, myself included, but with so little

success that he had not been able to dress himself for
five years : he could not lift the left arm, and natural
feeling was almost entirely gone from it.   The right
arm was also affected, but in a less degree, when he
applied to me on the 25th April, 1842.   I hypnotized
him, and he was so fully satisfied with the improvement he experienced, as to induce him to come to
Manchester to be operated on daily.   In a very short
time his improvement, both as regarded strength and
feeling, was most decided, as he could lift a heavy chair
with the worst arm, and could feel a small object such
as a pin, which could not have been distinguished by
him with that hand when I first saw him.   The pain
in his back was also speedily much relieved.   He was
exhibited at my conversazione to the British Association, 29th June, 1842, in this improved state, and has
made still farther progress since, although not yet able
to follow his usual avocation.   I should not omit to add,
that this patient was under my own care for some time in
1841, when, although he derived benefit from the means
used, he was not nearly so much or so rapidly relieved,
as by my present mode of treatment by hypnotism.

Case XVIII. Mr ——, 58 years of age, consulted me
in consequence of a paralytic affection of two and a half
years' standing.   Stated by his friends that he had had
an apoplectic seizure two years and a half before, which
was at first accompanied with total loss of consciousness, and of sense and motion of the right side for six
weeks.   He then gradually recovered, so as to be able
to walk a little in the course of four or five months.
When he called on me, 3rd June, 1842, his gait was
very feeble and insecure, always advancing the right
side foremost, his arm had always been supported in

a sling, he could raise it with an effort as high as the breast, had not the power of opening the hand, the thumb was much and rigidly flexed. Had little or no feeling in that hand. After being hypnotized for five minutes, feeling was restored, he could open the hand and grasp much firmer, and *raise it to his forehead.* His speech, which had been very imperfect, was also much improved. This patient was operated on for some time with partial improvement, so that he could manage his arm without a sling, and the feeling continued improved, and there was also slight improvement in his gait, but I was of opinion, that there was organic mischief in the brain which would prevent a perfect restoration, and therefore discontinued farther trials.

Case XIX. Miss Sarah Mellor had been under my care for nine months, for an affection of the lower part of the spine, accompanied with pain and weakness of the lower limbs, and with contraction of the knees, so that she had been unable to stand or walk without crutches during that period. I had used every means usually adopted in such cases, but instead of improving, she was getting worse in every respect, till I tried hypnotism, the satisfactory results of which were too immediate and apparent to admit of the slightest doubt of its great value on this occasion. The following is a statement attested by the patient :—

" Had suffered severe pain in my ankles, with contraction of the knees, and pain at the bottom of my back, so that I had been unable to walk without a pair of crutches for nine months. During this period, I had taken medicines internally, used liniments to the legs and spine, been leeched and blistered over the lower part of the spine, but still, instead of improving, I was getting worse, both as regarded the pain and contraction, so that I was becoming quite deformed, from the legs being bent on the thighs, and they on the body. I was thus about nine or ten inches less in stature than formerly, and than I am now. About the beginning of last March (1842) I came to Mr Braid, who had prescribed the other means to me without benefit, when he said he would try his *new method* with me. After being hypnotized *three times*, I was able to walk from my lodgings to the house of a friend who lived a few houses distant in the same street WITHOUT MY CRUTCHES, and in two days after, from that house to Mr Braid's WITHOUT CRUTCHES. I was operated on almost daily for three weeks, when I returned home, and at that time I was able to walk *half a mile without crutches.* After being at home five weeks, I returned to Manchester, and have been attended by Mr Braid for two months, and always found myself better after the operations. I took no medicine during my first stay in Manchester; and on this occasion having only done so when required for a violent cold on two occasions, from imprudent exposure. Since I came to Manchester last, one day I walked to Grosvenor Street, Piccadilly, and back again to my lodgings in Lower Mosley Street, fully a mile and a half, without inconvenience; on another occasion to Hulme and back again, *fully two miles. I was quite sensible, and could hear all that was said or done during all the operations.*

(Signed)   SARAH ANN MELLOR.
JANE LIVESEY, Witness.
C. WILSON, Witness."

MANCHESTER, 12th July, 1842.

This patient was exhibited at my conversazione 29th June, 1842. After returning home, she had the misfortune to get entangled by one of the feet in a cart rut, in a lane, which threw her back, but having returned and been hypnotized, I was enabled to send her home much improved, and when she called on me lately, she continued so.[1]

Case XX. Mrs J., 29 years of age, requested my attendance, 17th February, 1842. Had been attacked in the autumn of 1840, with slight degree of weakness of left side, and difficulty of speech, neither of which had ever been entirely removed. Three months after she was delivered of a still-born child, and had been affected with convulsions ten days prior to delivery, for which she seemed to have been treated in the usual manner. In about a month after delivery, 31st January, 1841, she had an apoplectic attack, attended with total loss of consciousness, and paralysis of the left side, for which her medical attendant had prescribed the usual treatment. I was called to attend her on the 17th February, and continued to do so for five weeks, when, as there was no particular improvement manifested, she passed into other hands, and after being under treatment with them for ten weeks, without improving, she was sent into the country, where she remained for about thirteen months, when she was brought back to town to be placed under my care, 15th June, 1842. The following was her condition at this period. Her mouth very much drawn to the right side ; her speech very imperfect ; and her mind confused. The left hand and arm were quite powerless, and rigidly fixed to the side, the hand clenched, the fingers and thumb being rigidly

[1] See Appendix II., Note 2.

and permanently flexed. The left leg very rigid, the heel drawn up, and the foot twisted so that it could only approach the ground by resting on the outer edge near the root of the little toe ; she could move this leg a little, but had never been able to stand, or walk a step, or support any weight on it. I hypnotized her, though owing to her mind being so confused, I experienced considerable difficulty in getting her to attend to the necessary instructions for producing the condition. However, I at length succeeded, and after the first operation—I kept her in the hypnotic state for ten minutes—she could hold her mouth much straighter, could move the fingers a little, and lift the hand and arm a few inches, and, with the assistance of her mother-in-law and myself supporting her by the arms, she was able to support half the weight of her body in walking across the room and back again. Her speech was also improved, and she evinced less confusion of mind. Next day I found the improvement was permanent, and hypnotized her again with advantage. 17th, Found her improved, and still more so after being again operated on. She could now, on merely steadying herself by laying hold of her mother-in-law's shoulder, stand supporting herself on the left leg, when the right foot was lifted clear from the floor. Her speech was still more improved, and mind more collected, so that I had very little difficulty in hypnotizing her now.

She was operated on daily, with advantage, till the end of that month, and the results shown to some of the most eminent professional and scientific gentlemen in this town. During the next two months she was operated on at times only, being so much better. In a few weeks she could walk to the door, steadying herself against the wall, and in a few weeks was able to walk into the street with the aid of a crutch. She had no medicine during this attendance. I only saw her occasionally now, and on the 17th September, when I had not seen her for nine days before, whilst taking her usual airing in the street, she was seized with apoplexy, from which she died within sixteen hours. On inspection, the whole of the superior and anterior lobes of the right side of the brain were found to be in a state of atrophy, only a thin layer, and that in a state of ramolissement, covering the ventricle, which was filled with serum, as was also the space between the pia mater and arachnoid, to make up the space vacated by the wasting of the cerebral substance. There was no effusion of blood. It is not at all surprising that such a case should have resisted former treatment, or proved fatal at last; but it seems surprising that, with such a state of brain, hypnotism should have had the power of producing so much improvement as it did.

Case XXI. 14th June, 1842, Mr Thomas Morris, 42 years of age, consulted me. He had had a paralytic stroke fifteen years previously, which deprived him entirely of the use of the right leg, and rendered the left weak and numb. In six weeks was able to walk a little, but never recovered entirely, being always weak and lame. Fifteen months ago had a second attack, with total loss of consciousness for a week, and also complete loss of voluntary power of the *whole body*. For several weeks required the urine to be drawn off by catheter. He has lately had the urine passing involuntarily sometimes, at other times voided with great difficulty. He has never regained the power of his legs so as to enable him to stand or walk without assist-

ance ; and has been, for the last six months, growing
worse. The arms very weak, being unable to raise the
right higher than the head, and even that accomplished
with great difficulty. Speech also very imperfect, and
his ideas so confused that he could make himself
understood with great difficulty. Hypnotized him for
five minutes, when he could speak much better ; could
raise his arm and hold an umbrella perpendicularly, or
horizontally, with his body, with perfect ease, and could
walk across the room, and back again, WITHOUT
ASSISTANCE, *for the first time since last seizure.*

(Signed)   THOS. MORRIS.

Witnessed by JOHN SHIPLEY,
Duncan Street, Strangeways.
C. C. MORRIS.
JOHN W. PACEY.
JAMES BRAID, Junior.

15th, Had the pleasure of finding the improvement
noted above was permanent, and also, that *he had been
able to retain his urine and void it at pleasure,* whereas it
had been passing *involuntarily,* both by night and day,
immediately before being hypnotized. He was again
hypnotized to-day with additional advantage. 17th,
Found him still better, having been able to walk in the
street with *one stick* for the *first* time for *last five years.*
Repeated the operation. 18th, Still better, so that,
with the aid of his two sticks, he had walked into
Ducie Street by himself. Operation repeated.

This patient went on improving, and on the 29th
June was exhibited at my conversazione. His speech
was greatly better immediately after *first* operation,
and his ideas seemed more vivid and clear. He was

also able to sign his name, and which he did very well, for the first time since his last seizure. Nor should I omit to add, that he had regained power over the rectum, which he had not previously ; and in about ten days he had got sufficient power of his hands to enable him to work. After he was considerably recovered he had the misfortune to fall, and injured the lower part of the back very much, which impaired the recently acquired power of the legs. They are somewhat better, but not nearly so well as they were a few weeks after he had been under my care. His arms, however, still retain their increased power, as I saw him lately lift a bed-room chair with the right arm, and hold it up nearly at full arm's length ; and the mind keeps pretty clear, much more so than before being hypnotized, notwithstanding he has had a severe attack of bowel complaint, from which he has been liable to suffer occasionally.

It would be difficult to adduce a more striking proof than the above, of the extraordinary power of hypnotism, there having been so many points at fault, all of which were immediately meliorated, and some of them permanently so.

Case XXII. Mr John W., 21 years of age, called to consult me, 18th April, 1842, for a paralytic state of the left side of the face, of thirteen days' standing. He had no power in the muscles of the left side of the face, consequently the mouth was drawn to the right, and he had no power of closing the left eyelid. In ten minutes after being hypnotized, and friction used, he could open and close the eyelid with facility, and had the power of retracting his mouth to the left of the mesial plane.

S

Case XXIII. 11th July, was consulted by Samuel Edwards, who had been unable to work for six weeks, in consequence of a paralytic state of the extensor muscles of the wrist, and a semi-paralytic state of the flexor and extensor muscles of the fingers. He had injured the arm by a heavy lift, and by a blow about two years before. The paralytic state came on suddenly about six weeks previously to my seeing him, accompanied by a tingling or prickling feeling in the fingers. I hypnotized him, calling into action the weak and entirely paralytic muscles in the best way I could. In consequence of this, he acquired the power of flexing and extending the wrist, when the arm was held horizontally with the ulna downwards, and of grasping pretty firmly with the fingers, immediately after the first operation, which he could not do before, as witnessed by several highly respectable individuals who were present the whole time. On the evening of the following day, he was able to milk a cow with his hand, and when he called on me two days after, I found him greatly improved. I operated on him again with additional advantage, and found him able to grasp so firmly that he could hold a single finger fast enough to enable him to be thus pulled from his seat without losing his hold.

He had undergone various treatment, including blistering, under two surgeons before I saw him.

17th July, 1842, he called on me, and had still greater power of the hand. After being again hypnotized, he could readily lift the one side of a heavy library table with the hand, which was quite powerless when I first saw him six days before. He stated, he had been able to work with it constantly from the time I saw him, on the 14th.

31st, He called on me, stated he had been improving. Was hypnotized once more.

August 7th, he called on me, and the first thing he did was to hold out his arm at full length, and show me he could bend and extend the wrist, whilst the arm was in the state of pronation. He had been able to do so for some days. Had been able to milk *five* cows the day previous. Hypnotized him again, after which he had still more power. He has not required to call on me since, being nine months ago. This patient must have continued well, as I have heard nothing more of him, which I was to do if he had any relapse.

I could easily multiply cases of successful practice in the treatment of paralysis by hypnotism, were it not for occupying too much space. I shall, therefore, condense a few.

XXIV. A gentleman sixty years of age had a paralytic stroke two years and a half before consulting me, which deprived him entirely of the use of the right arm, and enfeebled the right side and leg. When he called on me, he walked very feebly, could scarcely close the fingers and thumb, and could not extend them fully. He could with great difficulty raise the hand as high as the pit of the stomach, the pupil of the right eye was considerably larger than the left, and not quite circular; speech very imperfect. After being hypnotized for five minutes, he was able to open and close the hand freely, and to raise the hand above the head, and pass it to the back of the head, and he could also walk and speak much better. Pulse regular, —before operation, his pulse was very irregular. When he called on me next morning, I found the improvement had been permanent. I hypnotized him once more with advantage, and again on the two following days; seven weeks afterwards he called on me, when I found the improvement was permanent. He could speak and walk much better, could raise the arm, and move the fingers and hand freely, could pass the hand above and over the head, and take off his hat with it. The right pupil also was quite circular now, and nearly the same size as the other.

Case XXV. 4th June, 1842, Mr J. H., 67 years of age, had a paralytic stroke, 19 months previously, which deprived him entirely of speech, and of motion of right leg and arm. When he called on me, his speech was very imperfect, his hearing dull, and he had very little power in closing the hand, could raise the hand to the mouth, said he could sometimes raise it a *little* higher, but never so high as his head. After being hypnotized for five minutes, he could speak and hear much better, could grasp much stronger, and would raise the hand *a foot above the head*, and put his coat on without assistance, passing it over his head. His walking was also much firmer. He seemed greatly pleased with being able to put his coat on, as it was the first time since his seizure. He was also able to sign his name for the first time, to attest the accuracy of my report of his case, which he did before two witnesses who had been present during the operation. He called on me twice after this, the last time two weeks from his first visit, when I found the improvement was permanent.

Case XXVI. Thomas Johnstone, 36 years of age, had a paralytic seizure, 13th February, 1842, which deprived him of feeling and motion of left arm and hand. Had

partially recovered motion so as to be able occasionally
to move the fingers a little, and to raise the arm nearly
to the horizontal position, but frequently was suddenly
struck with pain and total loss of power of the arm, and
hand, and fingers, for four or five hours after. Had
been struck in this way just before I saw him, and he
was quite powerless, as above described, or rather the
arm was spasmodically fixed to the side ; had been
under medical treatment ever since his first seizure.
4th May, 1842, hypnotized him for four minutes, after
which he could move the fingers, hand, and arm freely,
elevating it above his head, across his body in either
direction, and could retain it in any situation he was
asked. The feeling, however, was still very imperfect.
5th May, called on me to go to my lecture, when he
had the complete control of the hand, arm, and fingers.
He was hypnotized in the lecture-room the same night,
and in four days after, the feeling, as well as power, was
restored to it. 26th, called on me again, and has perfect voluntary power of the arm, as well as natural
feeling and heat of the member. Attested as correct
by the patient.

(Signed by proxy to which the patient affixed his
mark) THOMAS JOHNSTONE.
Witnessed by JOHN HARDING.

I have also a copy of a certificate of his condition
from the physician who attended him immediately
before he consulted me. On the 20th January, 1843,
his father informed me that his son had requested him
to call on me, and say he was in America, and had
remained well ever since I saw him, and, that he wished
his father to express how grateful he felt for the benefit
he had derived from my operations.

I shall only give one more case illustrative of this class.

Case XXVII. Mr H., 39 years of age, had been partially paralytic of the inferior extremities for ten years, which came on some time after a fall, accompanied with double vision. The latter disappeared under treatment, but the former increased. When he called on me, 18th February, 1843, he was walking with a crutch and stick, and with the assistance of both and a servant, it was with great difficulty he could ascend the few steps at my door. After the first operation, he could walk across the room and back again, *without either crutch or stick*, and after being operated on next day, he was able to mount twenty-eight steps to his bed-room, without his crutch, and has done so ever since. In ten days, I was agreeably surprised to see him on the fourth bench of the lecture-room of the Manchester Athenæum, to which he had ascended eighty-one steps, with the aid of a stick in one hand.

This patient had not been aware, until I called his attention to the fact, that he had very defective vision of the right eye, and was surprised to find on testing this, that when the left eye was closed, he could with difficulty see the large heading of the leader of the *Medical Gazette*, whereas he could read the ordinary sized print of that article with the left. After being operated on, he could read the small print of the leader with the right eye also, at which he felt greatly surprised, as well as at the increased power of his legs, because, as he had been conscious all the time of the operation, he could not believe I had done any thing to him, till he found on trial he had been so much benefited in both functions.

Here, then, we have the beneficial results most unequivocally ensuing even when the patient imagined no effect could have been induced. The improvement in the sight has remained permanent, and he also improved in the power of his limbs, till he had the misfortune to fall, while carelessly looking at something when walking on the street one day.

In confirmation of the efficacy of a few minutes of hypnotism, in curing many cases of paralysis, I may refer to the reports of the Liverpool papers, as to what took place at my lectures in that town in April, 1842. There were hundreds who witnessed the effects when I publicly operated on such patients, who were entire strangers to me. Cases where the patients had been for years powerless of limbs, so that they could not unlock the clenched hands, nor raise the arm to the chin, even with the aid of the other arm, have been enabled in eight or ten minutes, to open the hand, and lift the arm above the head. My intelligent friend, Mr Gardom, lately informed me, he had treated a paralytic case most successfully by hypnotism.

Case XXVIIᴬ. Mrs E., thirty-seven years of age, had a paralytic affection when thirteen months old, which deprived her entirely of the use of the right leg, which has never been recovered. At seven years of age, she had a second attack, which deprived her also of the use of the right arm, which was recovered after nine months' professional attention to it. At fifteen years of age she had a third attack, which drew her face, and deprived her of speech for some time, but was recovered from ; and she had no farther attack of the sort till 8th January, 1842 (being twenty-two years from former

attack).   The latter attack enfeebled the right arm,
and completely paralyzed the whole of the *left
side*.

Being of full habit, she was bled from the arm, had
active cathartics, leeches, and blisters.   In six days
there was improvement to this extent, that the right
hand could be raised as high as the shoulder, the left
arm could be moved feebly, and the hand closed feebly
and slowly.   When sitting on a chair, the left leg could
be moved with great difficulty, so as to raise the heel
from the ground.   I hypnotized her, and in five or six
minutes she could raise her right hand and arm *above
her head*, could move the left arm freely, and grasp
firmly, and could raise the left leg so as to place the
heel eighteen inches from the ground.   Next day she
was able to walk across the floor with her one crutch.
A pain in the knee induced her to avoid walking afterwards, but in three weeks she could walk quite cleverly
as before last attack.

The other cases were all in the chronic state, of long
standing, and had resisted all ordinary means, and the
restorative powers of nature and time, and yet we have
seen what extraordinary powers can be exerted, and
effects produced, in such cases by hypnotism.   The
latter proves its superior efficacy, to other means, in
more recent cases.

Case XXVIII.   I shall conclude the subject of paralysis with the following most interesting case.   The
subject of it was Miss Atkinson, a middle-aged and
very intelligent lady, and I shall give the case as
recorded by herself in a letter she was so obliging as
to furnish me with, for the purpose of publication in
this work.

RATIONALE OF NERVOUS SLEEP   281

LETTER FROM MISS E. ATKINSON
(*of the Priory, Lincoln*).

"MOSLEY ARMS, MANCHESTER,
Monday, 4th July, 1842.

"Dear Sir,—I have very great pleasure in furnishing you with a statement of my case. I beg you will make whatever use of it you think proper, and most sincerely do I wish that it may lead others suffering from disorders on the nerves, to seek relief from the same source, and with the same success.

"In January, 1838, I was attacked with cold and influenza, accompanied by a violent cough, on the 29th of this month. Ten or twelve days after the first attack, without any previous warning, my voice left me instantaneously, and I could not utter a sound louder than the faintest whisper. For three weeks I had no medical advice, hoping daily, from my ignorance of the nature of the complaint, that my voice would return; but being disappointed, and feeling my health and strength declining, I consulted Mr Howitt, an experienced and eminent surgeon in Lincoln, who immediately requested I would confine myself to my own lodging-room, which was to be kept at a regular temperature. He prescribed such medicines as my case required, and ordered blisters to my throat and chest, which were kept open, until I became so completely debilitated that it was considered necessary to discontinue them. Towards the latter end of April my health was considerably improved, and I was allowed to leave my room, though my voice was still merely a feeble whisper. Shortly afterwards, I paid a visit to a sister in York, whose family surgeon, Mr Caleb Williams, a man in extensive practice, prescribed for me, and took great

interest in my case.   Soon after my return to Lincoln,
I consulted Mr Joseph Swan, 6 Tavistock Square,
London, who entirely approved of the treatment I had
undergone, and prescribed such additional remedies
and medicines as he thought would be beneficial.   Since
then he has continued to visit me whenever he has been
in the country.   Galvanism has been tried without producing any effect; electro-magnetism also, by a scientific friend (not a medical man).   I have frequently
conversed with several other professional gentlemen,
who have also taken a great interest in my case.   They
all agree in opinion that the attack was paralysis of the
organs of voice, without disease; and that the treatment I have undergone has been most judicious; in
fact, that every thing has been done for me the medical
profession could suggest.   Every one of them has told
me, that when my health and strength returned, there
was every reason to believe I should recover my voice.
I remained in a very weak and delicate state for some
time, but have now been in perfect health for more
than twelve months, yet without having the power of
speaking above a whisper.

"I considered the recovery of my voice hopeless,
until hearing of the many cures you had performed by
hypnotism, I was induced to state my case to you, and
request your opinion as to the probability of this system
benefiting me.   Your reply was, ' *If*, as seems to be the
opinion of most of the professional gentlemen consulted,
your loss of voice is owing to *exhaustion* of the *nervous
energy* of the *vocal nerves*, and not to *positive destruction
of any portion of them*, I consider my mode of operating
is likely to be very speedily successful.   On the other
hand, if there is positive destruction of the nervous

substance, *with loss of continuity of the principal trunks of the nerves*, it will alter the chances very materially. However, as this cannot be positively known without trial, and as the extraordinary power we possess of rousing nervous energy may be sufficient to enable the function to be restored with a state of nerve which could not be of service under any other agency, I should decidedly give it as my opinion that it ought to be tried, as no risk can attach to the trial, and a week or two at most, will be all the time required for giving it a fair trial.' This raised my hopes; I came to Manchester on Tuesday the 28th of June. You operated on me twice that evening, and twice each succeeding day, but without producing any change on my voice until Saturday, July 2nd, when, on rousing me from the hypnotic state, I spoke aloud without the slightest effort. My voice was then weak ; you have continued to operate on me until now, Monday morning (4th July), and my voice is fully restored to its original strength, with the power to vary its tone at will. Thus has hypnotism given me back the power to make myself understood by those to whom I address myself, of which I had been deprived for the last four and a half years. I have not suffered the slightest pain or inconvenience while submitting to the operations, nor any unpleasant effects afterwards ; neither did I ever once lose consciousness of all that was passing around me.

" With heartfelt humble thanks to our heavenly Father for this and every blessing, particularly for the hitherto unknown power bestowed on man ; and with deep gratitude to you for your kind attentive care while so skilfully and successfully using this power for the restoration of my voice, I beg you to believe

me, dear sir, yours very respectfully, and greatly
obliged,

"ELIZABETH ATKINSON."

It is but justice to the professional gentlemen who
had been consulted in this case prior to application
being made to me, to say, that I consider they had
treated the case most judiciously, according to our
previous experience in such cases; and it must be
interesting to them to find that in this agency our
art has acquired a new and efficient resource for such
cases.

This case is interesting in many points of view. The
circumstance of the patient having been operated on
twice each day successively, that is, *eight* times, without
any visible improvement—for I had her tested before
and after *each* operation—and being able to speak
aloud, without effort, on being roused from the hypnotic
condition on the *fifth* day, is sufficient proof that the
improvement was *not the effect of imagination*, but of
the physical condition induced by carrying the operation farther. Any effect to have been anticipated from
mere mental emotion, we should have expected to
have been greatest at first, and to have become less
and less as the party became familiar with such operations. Here, however, it was quite the reverse. I
found, on testing the patient on the 2nd of July, *immediately before being operated on*, that no improvement
had been effected from the former operations (she had
been operated on eight times), and therefore resolved
to carry it farther that time; and the result was, as
noted above, that on being aroused she spoke aloud
without effort. It is also important, as corroborating

the statement of many others who have been cured of various obstinate complaints by hypnotism, that they *could hear quite distinctly, and retained consciousness the whole time*, of all that was going on around them. In some cases, however, it is necessary to carry it to the ulterior stage, or that of *insensibility*.

On the 19th October, 1842, I had the pleasure of receiving a letter from Miss E. Atkinson, from which I make the following extract, in proof of the *permanency* of the cure. "You will be glad to hear that I have retained my voice without any intermission, since I left you. The only difference is, that it has become stronger; and my health is in every respect perfectly good." I had also the pleasure of hearing from a friend, a few days ago, that she still continues well, and it is now nine months and a half since her voice was restored.[1]

I doubt not there may be some who, on reading the cases I have recorded in this treatise, will be disposed to appeal to the well-known fact, that various complaints have been suddenly cured by mere mental emotion, hoping thus to throw discredit on the curative powers of hypnotism. Whilst I grant the premises, I deny the justness of the inference. That I may meet the subject fairly, I shall now quote some of the most remarkable cases of the sort which have been recorded. "Dr Gregory was accustomed to relate the case of a naval officer, who had been for some time laid up in his cabin, and entirely unable to move, from a violent attack of gout, when notice was brought to him that the vessel was on fire ; in a few minutes he was on deck, and the most active man in the ship. Cases of a

[1] See Appendix II., Note 3.

still more astonishing kind are on record. A woman, mentioned by Diemerbroeck, who had been many years paralytic, recovered the use of her limbs when she was very much terrified during a thunder-storm, and was making violent efforts to escape from a chamber in which she had been left alone. A man, affected in the same manner, recovered as suddenly when his house was on fire ; and another, who 'had been ill for six years, recovered the use of his paralytic limbs during a violent paroxysm of anger." Abercrombie *On the Intellectual Powers*, pp. 398-9.

To these might have been added the influence of the sight of a tooth key or forceps, or even the approach to the house of a dentist, in curing toothache.

Now, what are the legitimate conclusions to be drawn from the history of such cases ? Is it not simply this, that such results *are possible*, and that they can be effected by *different means?* Now, as it is apparent that analogous results can be induced by hypnotism, I would ask is hypnotism not quite as convenient and desirable a remedy as setting a ship on fire, raising a thunder-storm, converting the patient's house into a bonfire, or exciting him into " a violent paroxysm of anger ? "

Again, of those who talk so much about the power of imagination, I would ask, what is it? How does imagination act to produce such extraordinary and contradictory results? For example, the mental emotions of joy and sorrow, love and hatred, fear and courage, benevolence and anger, may *all* arise either from *real*, or from *imaginary causes only*, and may seriously affect the physical frame. In many instances these different and opposite emotions have proved almost instantly fatal; in other instances equally sanative. How is this achieved? Are not the whole of the emotions accompanied by remarkable physical changes, in respect to the respiration and circulation as well as sensation? Are they not highly excited in one class, and depressed in the other? And may not this act as the proximate cause in effecting the permanently beneficial results during hypnotism? As already explained, analogous physical results can be produced by hypnotism; and it is no valid reason why we should not profit by it in the treatment of disease, that we cannot positively decide as to its *modus operandi.* It seems quite evident that we have acquired, in hypnotism, a more ready and certain control over the physical manifestations referred to, and which can be turned to useful purposes, than by any mode of acting on the imagination only, which has hitherto been devised.[1]

Rheumatism is another affection, for the relief of which I have found hypnotism a most valuable remedy. I have met with some cases of rheumatism, however, which have resisted this, as they had every other method tried; and others, where it afforded only *temporary* relief; but I am warranted in saying, that I have, on the whole, seen far more success, more rapid and decided relief, follow this mode of treatment, than any other. It has been chiefly in chronic cases in which I have tried it. In its application, I first induce the somnolent state, and then call into action the different muscles which I consider directly affected, or which, by being so called into action, are calculated to change the capillary circulation and nervous sensi-

[1] See Appendix I., Note 14.

bility of the part implicated. The patient must be retained in such position a longer or shorter time, according to circumstances. The following cases will illustrate the effects of this mode of treatment:—

Case XXIX. Joseph Barnet, near Hope Inn, Heaton Norris, Stockport, 62 years of age, called to consult me on the 10th December, 1841, for a severe rheumatic affection of the back, hip, and leg, of thirteen years' standing, which had been so severe, that he had not been able to earn a day's wages during that period. He had been equally a stranger to comfort by day, as to refreshing sleep by night. He came to me leaning feebly over his stick, suffering anguish at every step, or movement of his body. He was treated at the commencement of his complaint by a surgeon; but feeling no relief, like many others similarly afflicted, he had recourse to all sorts of nostrums, and also to hot salt water baths. I hypnotized him, placing him in such attitudes as his particular case required, and in fifteen minutes aroused him, when he was able to bend his body freely, and not only to walk, but even to run. He called on me in a few days after, when he stated he had slept comfortably, and been perfectly easy from the time he left me till the night before. I hypnotized him again with advantage, and a few more times sufficed to restore him entirely. This patient was seen, and bore testimony to these facts, at two of my lectures. After one of them, from being too late for the coach, he walked home, a distance of six miles. This was by no means judicious, but proves incontestably his great improvement.

I was not at that time so well aware, as I have been since, of the great power of hypnotism in such cases, and therefore ordered him some medicine after the first operations; but from observing that the relief immediately followed the operation *before taking medicine*, and that the pain returned in some degree the night *before next visit*, and when, had there been benefit resulting *from the medicine, it ought to have been diminished after using it*, and that relief was again afforded during the hypnotism, I felt convinced the *medicine* had *no share in the improvement*, and therefore discontinued it, and trusted entirely to hypnotism. In the beginning of January, 1842, when this patient called on me, he was so well, that I told him farther operations would be unnecessary for the present, but added, that should he have any relapse, if he called on me again, I would hypnotize him, without charge, of which offer he promised to avail himself.

At my lecture on the 27th December, 1841, several questions were put which elicited the following answers :—
" Do you mean to say you were never so well as you are now ? "—" Yes ; I never earned two shillings during all that time.   This last winter I was worse than ever."
" Did you walk, sir, before ever you left my surgery, without taking any medicine ? "—" I did, and ran too."
See *Manchester Guardian*, 1st January, 1842.

I heard nothing farther of this patient for about seven months, and therefore, after the offer I had made him at last visit, had every reason to conclude he had remained well.   However, it appears he had a relapse shortly after he left me, and his family, upon whose exertions he depended, being out of work, he could not afford to pay the railway charge for coming to see me again. His relapse having been laid hold of, and construed into a charge against me as having falsely represented

T

his case, I was induced to call on the patient, accompanied by two friends, when he furnished us with the following document :—

"Joseph Barnet, Providence Street, Heaton Norris, had suffered from a severe rheumatic affection, prior to last December, when he applied to Mr Braid. He was first under the care of Mr ——, Higher Hillgate, who bled, blistered, and prescribed medicines for him; but the complaint remained unabated. From this period, took various medicines, and other means recommended to him by those who had been similarly afflicted, and who considered he would be benefited by such means as had relieved them, but received no relief. After that applied to Mr —— of Manchester, from whom he considered he derived benefit for a fortnight; but the pain returning, he went to Liverpool to the warm baths, where he remained as long as money lasted, but without being relieved.

"From this time tried various means as recommended by different parties. During the whole of this period he had never been able to earn a day's wages. When he applied to Mr Braid in December last (1841), had been suffering extreme pain in every movement of the body; in short, he had walked nearly double, supported on a stick. He was operated on by Mr Braid, and in a quarter of an hour he was roused, and found himself able to walk and run. At first, Mr Braid walked him about by the hand, and afterwards made him run without any assistance whatever, as his wife and others present can testify. The case as stated by Mr Braid in his lectures in my (his) presence was perfectly correct, as I (he) bore testimony to at the time. Owing to being unable to pay the expenses of the railway, he did not return to Mr Braid, when he had a recurrence of the pain. He had never informed Mr Braid, that he had had a recurrence of the pain, and never saw him afterwards until the evening of the 26th July, 1842.

his

(Signed)   JOSEPH + BARNET.

mark

J. A. WALKER.

THOMAS BROWN.

HARAIT BROOKS."

(Daughter of J. BARNET.)

Case XXX. 11th January, 1840. Mrs B., 48 years of age. Catamenia ceased last spring. Has suffered from a severe rheumatic affection for the last three months, and been confined for the last two months to her bed-room. The legs, arms, neck, and head, were excessively painful, so that the slightest movement was attended with great agony. She was quite alarmed at my taking hold of her arm to feel the pulse. When in bed could not turn over, nor bear the slightest touch. 11th January, 1842, hypnotized her, and roused her in ten minutes, when she was quite free from pain, being able to walk, stoop, and move the arms, wrists, and fingers, with perfect freedom. 12th, had slept comfortably all night; had been able to lie on her side, which she could not do before for three months; could rise from the chair, and move legs and arms without pain. There was, however, a soreness or uneasy feeling, although not amounting to pain, in some parts of the limbs. Hypnotized her for eight minutes, when she felt less of the numbness, and followed me down stairs, and ascended them again, without taking hold of the banister, and taking the steps regularly and cleverly with both feet alternately. 14th, Found her down stairs enjoying herself with her father, husband, and friends, almost quite well. Hypnotized her again, and also in a day or two after, and she had no recurrence of the rheumatism, although a degree of stiffness of the limbs remained. She had no medicine from me till the rheumatism was gone, when she had some for a different complaint. This patient was seen at my house seven months after by about sixty friends, including several professional gentlemen, when the above statement was read in her presence, and confirmed by her as correct to that time ; and as I have heard no intimation, I feel assured she has not had a relapse.

Case XXXI. Mrs S. has been already referred to, Case VI. She had suffered much from rheumatism for many years, and had never been entirely free of it, notwithstanding she had undergone much treatment. After first operation she was much relieved, and after a few more was entirely free from pain. It has recurred occasionally since, but has always been removed by one or two more operations of the same sort, and which are neither painful, nor in any way unpleasant.

Case XXXII. Another rheumatic case of a patient 53 years old, of seven years' standing, where sleep had not only been courted by exhausted nature, but also by the most powerful doses of narcotic drugs ; on one occasion 400 drops of laudanum had been taken in *two hours* ; still the pains continued, and yet, by *fifteen minutes* of hypnotic sleep, procured by the simple agency I recommend, this patient was relieved of his agonizing pains. In this case, from my knowledge of the eminence of the professional gentleman who had prescribed for him, I feel assured every known remedy had been resorted to, but without effect, and yet this agency succeeded in a few minutes. This patient had suffered severely for seven years ; was first hypnotized 10th February, 1842, and again on the 17th and 19th. He seemed as nearly as possible entirely free from pain, and had suffered very little after the first operation, less than at any previous period during the seven years he had been a rheumatic subject. I have lately heard he had a relapse some time after I last saw him ; but no reasonable person could expect three operations should have sufficed to eradicate such an obstinate complaint permanently; most probably a repetition of the process would.

Case XXXIII. Mr John Thomas, 155 Deansgate, consulted me at the end of April, 1842, for a severe rheumatic affection of the loins, and right hip and leg, which had continued for two weeks. Had a rheumatic fever two years before, which confined him to bed for sixteen days, and to his room for a week longer ; and he did not get rid of the pains for three months after he was able to go out, although he tried Buxton and Matlock baths, and also the medicated and sulphur baths in Manchester. When he called on me (April, 1842), I hypnotized him, and when roused he was almost entirely free from pain, and never required a repetition of the operation. He had no medicine. On the 28th July, he called on me to say he had continued quite well in every respect from the time he was hypnotized, and attested the same, and the correctness of the above statement, by appending his name to it in my case-book, and he has also been seen by many professional and other friends who can bear testimony to the same effect. He continued well when I saw him lately.

Case XXXIV. Master J. Lancashire, 12 years of age, was brought to me in September, 1842. He was suffering from a violent rheumatic affection of the legs, back, and chest, so that he required to be carried into my house. After being hypnotized, he was so much relieved as to be able to walk about the room freely, and to walk to his cab without assistance. Next day he called, and was hypnotized again, and left my house quite free from pain, and has kept so well as never to require another operation. He had no medicine, either externally or internally. His mother and he called some time after to inform me he had remained quite well, when they both attested the correctness of the above statement of his case.

Case XXXV. Mrs P., a lady upwards of 50 years of age, had suffered so severely from rheumatism that she had not enjoyed a sound night's sleep for seven months. External and internal means, which had been beneficial in a former similar attack, had been tried without effect, before I was sent for to visit her. She was suffering excruciating pain in one leg, particularly about the knee joint. When I proposed to relieve her by hypnotism, she repudiated the idea, told me she had no faith in it, and felt assured in her own mind such an operation could be of no use to her. I told her I cared little for her want of faith in the remedy, provided she would submit to be operated on as I should direct. She at last consented, and in the presence of her three daughters was hypnotized. In eight minutes she was aroused, and was quite free from pain ; wished to know what I had done to her ; said she felt assured hypnotizing her could not have relieved her. To this I replied by asking where her pain was felt now. She answered she felt no pain, but persisted she was sure I had done nothing to take it away. The manner in which she could walk and move her limbs was sufficient proof the pain was gone, notwithstanding her scepticism about the agency. When I called next day, I was informed by her family *she had slept comfortably all night*, and had gone out, being quite well. Two days after, I called again, and was informed by her that she had been overtaken in a shower, and had over-exerted herself on that occasion, and had had a return of the pain, although not so bad as at first. I hypnotized her again with complete relief, and she has never required a repetition of the operation since, so that she has now enjoyed a release from her old enemy for eleven months, in defiance of her scepticism. Here, then, we have a very decided proof that it was not imagination; in short, that it was a physical and not a mental change which effected the cure.

Case XXXVI. Mr Hampson, another rheumatic case, I was called to 16th May, 1842. The patient was a powerful young man, 23 years of age; had suffered severely for three weeks, the last two been entirely confined to bed, unable to move his legs, or to feed himself; for two weeks had not known what it was to have ten minutes continuous sleep, from the violence of the pain, and spasmodic twitching of the limbs rousing him. His left hand, fingers, and wrist were so swollen and painful, that he was quite alarmed at my attempting to feel his pulse. After being hypnotized for five minutes whilst in the recumbent posture, I had his arms extended, and he was now roused and able to

move the wrist and fingers with comparative ease. I
now hypnotized him once more, and operated on his
legs. In *six minutes* he was able to get on his feet,
walk round the bed and back again, and get into bed
and lie down *without assistance.* Next morning I found
him up and dressed, and able to walk very comfortably.
He had slept well through the night. I hypnotized
him again. Next night he slept uninterruptedly, and
in the morning felt nothing of his pains excepting in
the left shoulder ; but this was quite well by the next
day. He had no medicine except a mild aperient.

The cases adduced I consider sufficient to prove this
to be a valuable agency in the treatment of *chronic*
rheumatism. I shall now adduce the results of its application in two cases of *acute* rheumatism.

Case XXXVII. Mr G., a literary gentleman, consulted me last winter. I found him complaining of
severe pain in the right arm and hand ; one point, the
size of a crown piece, on the outer edge of the arm, a
little below the elbow joint, was exquisitely painful.
He was enveloped in double clothing, but, notwithstanding, was quite starved and chilly with cutis
anserina, pulse 120 strokes a minute. I told him I
considered it was the commencement of an attack of
rheumatic fever, and I should wish to try whether it
could be cut short by hypnotizing him. He had never
been operated on in this way before, but readily assented. In six minutes I had him bathed in perspiration, and his pain greatly relieved. He was now
ordered to bed, to take a mixture with vinum colchici.
Next morning I found him much freer from pain, it
had never been severe since the operation the day
before, the skin comfortable, and his pulse only 80.

To remain in bed and continue his medicine. Next day the pulse was 70, and no complaint of pain, and the following day he was able to go out and attend to his business. No relapse.

Case XXXVIII. Mrs B., the mother of a numerous family, had a severe attack of rheumatic fever, affecting different joints in succession, and also violent pain in her head. I proposed she should be brought out of bed and hypnotized. The pain of her knees, feet, and ankles, was so severe that she could not stretch her legs, nor attempt to support herself, in the least degree, upon them. She had therefore to be carried from the bed to the chair where she was to be hypnotized. In five minutes she was roused, the headache gone, and the pain in her legs and feet so much relieved that she was able to walk to bed, requiring only to be slightly supported by the arm. The pains never returned with the same degree of severity. She was hypnotized a few times more, and always with benefit. Of course I prescribed such medicines as I considered necessary to improve the state of the secretions, so as to put as speedy a termination to the attack as possible, but there could be no doubt that hypnotism contributed very much to meliorate her suffering, and also in bringing the attack to a more speedy termination, than would have been the case had I trusted to the effects of medicine only.[1]

The following cases can perhaps scarcely be introduced in any other place with more propriety than the present. They are cases of painful affection of the members, arising from irregular action of the muscles, consequent on mechanical injury.

[1] See Appendix II., Note 4.

Case XXXIX. Mr J. J. consulted me on the 6th
November, 1842. He stated he had a fall from a
horse five months previously, when he sustained severe
injury of the left hip and thigh. He was confined to
bed for two weeks, under medical treatment, supposing
the parts to be only bruised and sprained. He then
began to move about with crutches, but with great
pain; and a consultation being held, it was considered
there was dislocation of the hip joint, but the attempts
made to reduce it failed. At the end of nine weeks
from the accident, another surgeon, 40 miles off, was
sent for, who confirmed the opinion that there was
dislocation of the hip joint, and he succeeded in reducing it. The patient was now confined to bed for
two weeks, and, on rising, was able to move about with
the aid of a stick, but without crutches. However, he
was still very lame, and in much pain. When he
called on me, which was on the 6th November, 1842,
he was not suffering much pain, but was extremely
lame. The knee was a little advanced forwards, and
the toes considerably everted. In attempting to walk
without the aid and support of his stick, the body was
thrown so much to the left at every step, as if the leg
were considerably shorter, that with other circumstances
coupled with this, led me to suspect fracture of the
neck of the femur within the capsular ligament. A
minute examination satisfied me this was not the case;
and I now considered the affection was one of irregular
action of the whole muscles of the hip and thigh, some
being atrophied and semi-paralyzed, and others inordinately tense. With this view I believed I should
be able to rectify the irregular distribution of nervous
and muscular energy by hypnotism, an opinion the

correctness of which was quickly verified. Having hypnotized the patient, and placed the leg in that position calculated to restore the functions according to the view I had taken, in about six minutes he was roused, and was agreeably surprised with such a remarkable improvement. Next morning he was again operated on, and was then almost entirely free from lameness, and entirely free from pain, so that he asked my opinion whether I considered it at all necessary for him to take his stick in going through the town on some business. He called on me the three following days, after which he went home, equally gratified as myself with the result of our operations. He had no internal medicine, nor external application, whilst under my care. He attested the accuracy of the above report before leaving ; and, as I have not heard from him since, have reason to believe he continues well.

This patient was seen by several gentlemen, some of them members of the profession, who can bear testimony to the correctness of these statements, as they had an opportunity of hearing the whole from the patient himself.

Case XL. Mr J. H., 68 years of age, called to consult me on the 8th November, 1842, relative to a painful state of his left shoulder, the consequence of a blow he had sustained two months previously. He had been under the care of two eminent professional gentlemen from the time he received the injury till within a few days before I saw him. There was a wasting of the muscles about the shoulder, great pain in moving the arm, and it was so weak that he had not been able even to button his coat with it. After being hypnotized the first time he could use it, raising it above his head, and

moving it in any direction with ease and freedom.
After being operated on next day he had still more
power. The following day he felt a little pain behind
the shoulder, under the scapula, which was entirely
removed by being once more hypnotized, and calling
the affected muscles into action. On Saturday, the
12th November, 1842, he left me, quite well, to return
home to attend to his business. Both this patient and
his son attested the correctness of the above report in
my case-book.

Case XLI. J. W., 25 years of age, had a severe
injury of the hip, which was followed by suppuration
between the trochanter and ischium, where there was
a fistulous opening; the leg was flexed and perfectly
useless, being supported by a sling passed over his
shoulders, whilst he supported himself very feebly on
two crutches, his health having suffered greatly during
his affliction. He stated that he had just left a public
institution, where he was given to understand no hopes
were entertained of his recovery. I hypnotized him,
and during that condition regulated the malposition
of the limb, stretching the contracted muscles, and
strengthening others, by exciting into action those
which had been weakened by being overstretched and
enfeebled by inaction. The result was, that on being
aroused he could straighten his leg, and walked (using
his crutches of course) with the sole of his foot resting
on the floor. He was operated on daily with the most
marked improvement both as regarded his leg and his
general health. In three weeks he could walk with one
crutch, in two weeks more threw that aside, and walked
with a stick, and shortly after could walk without that
aid, and is now well, excepting a little weakness of the

ankle joint. He had no internal medicine from me, and no external application, excepting one box of ointment, the discharge having entirely ceased within a week of his being under my care.

I shall now advert to the remarkable power of this agency in speedily overcoming nervous headache. I have so many examples of this, sometimes two or three fresh cases in a day, that it is almost useless to instance individual cases. However, I shall give a few.

Case XLII. Mrs B., the mother of a family, has been constantly annoyed with headache and maziness, for the last two or three years, varying in intensity at different times, but never entirely free from it. Consulted me, 22nd January, 1842, for the above complaints, and also stated that she was subject to attacks of epilepsy. I hypnotized her, and in five or six minutes aroused her, when she was quite free from headache. She was hypnotized almost daily for some time, and remained quite free from headache, five weeks after she was first operated on, and had much less of the mazy feeling, and no fit for two months. She appeared so much better as to be taken notice of by all her friends.

Case XLIII. Miss B., daughter of the above, was brought to me on the 23rd January, 1843, in consequence of the improvement her mother had experienced from the operation. She had suffered severely from headache for six months, so much so, as frequently to cause her cry and shed tears, and was never entirely free from it for that period. I hypnotized her, and in five or six minutes roused her quite free from headache or any other ache. She was operated on almost daily for some time, and has had no return of the headache to this time,—four months,—and has had her appetite

much improved, and looks very much better.   She had
no medicine.

Case XLIV. Miss S., on the 25th January, 1843,
was suffering from a most violent headache, and had
been so all day.   She could scarcely open her eyes or
see when they were open, and seemed quite prostrated.
I hypnotized her, and in five minutes she was aroused
quite well, and has had no return of it at the end of
ten days.

Case XLV. Miss N., 20 years of age, had suffered
severely from headache from childhood, and never
knew what it was to be entirely free from that complaint, but frequently had it so severely as to incapacitate
her for any exertion, and almost to deprive her of sight.
She also had constant uneasiness at stomach, sometimes
amounting to severe pain, and when the attacks of
headache were at the worst, the pain at stomach was
also much aggravated, and a severe attack of vomiting
generally terminated the violence of these paroxysms.
In April, 1842, I hypnotized her, and from that period
she has been almost entirely free from both headache
and stomach complaint.   At the end of fifty-four weeks,
I had the pleasure of hearing from herself, as I had
previously from her mother, that she scarcely had
suffered from headache at all since the operation, and
never severely, or even in the slightest degree for one
hour at a time.

Case XLVI. Mrs T. had been suffering from severe
pain of the head for more than two weeks, without
intermission either by night or day when awake.   She
had also had severe pain of the left side of the chest
for three weeks, which was aggravated by a cough.
For the last two days, the pain of the side had been

most distressing. The pulse was rapid, the cough frequent and severe, and the pain in the side so acute as
to prevent free expansion of the chest as in ordinary
respiration.   I found there was considerable spinal
tenderness on pressing betwixt the shoulder blades.   I
hypnotized her, and in five minutes, when aroused, she
was quite free from headache, the pain in the side so
much relieved, that she could move her body freely,
and take a moderate breath with very little inconvenience.   Next day I found she had no return of the
headache, and very little of the pain in the side.   She
was again hypnotized with advantage, which I repeated
daily, and in six days the pain of the side was quite
gone, the pain of the head had never returned, the
cough was gone, the spinal tenderness which disappeared at first operation had never returned after the
first operation, and the patient was now quite con-
valescent.   She had no medicine but some pectoral
mixture to moderate the cough.[1]

I shall now refer to spinal irritation, which is well
known to be the source of much suffering, not merely
in the course of the spinal column, but also, from its
influence on the origins of sentient nerves, on distant
parts of the body.   I have already referred to this in
the cases 16 and 19, where there was loss of feeling
and motion in one case, and pain of the legs with
contraction in the other.   Where the affection does
not depend on active inflammation, I hesitate not to
say, that the pain of the spine, and other painful
affections dependent on the state of the spinal nerves
which arise therefrom, may be relieved more speedily,
and certainly, and effectively, by hypnotism, than by

[1] See Appendix II., Note 5.

any means I have either tried, read, or heard of. I shall give an example or two.

Case XLVII. Miss C. had suffered for years from spinal irritation and headache, the pain extending round the chest, so that deep breathing or free motion of the chest could not be tolerated. I tried every variety of treatment, but in vain, and at last despaired of benefiting her, and, from the extreme difficulty of breathing, suspected it must end in pulmonary consumption. I now tried hypnotism, which immediately succeeded in relieving the whole catalogue of painful symptoms, and she was speedily restored to perfect health, and has continued so ever since.

Case XLVIII. Miss —— had suffered much from spinal irritation for years, and had undergone much severe treatment. Had been restored to health and strength under my treatment, but was again threatened with a relapse. I hypnotized her, and when roused, the spinal tenderness was gone. A few more operations made a most marked improvement, and she continued well for some months. She had a recurrence of the complaint, when hypnotism was again had recourse to, with immediate and decided advantage.

I could easily multiply cases of this sort, were it not for swelling the volume unnecessarily. I shall therefore pass on to cases of irregular or spasmodic action of the muscles. I have found it decidedly useful in several cases of chorea ; and also in cases of nervous stammer. In epilepsy it also frequently proves highly useful, but there are some varieties of this complaint over which it has no control. These I presume are such cases as depend on organic causes, and which are found to resist every known remedy. It is however well known that many cases which were supposed to have been of this class have worn themselves out, or time and the efforts of nature have effected some organic change. Whether hypnotism, if persevered in, might have a tendency to expedite the favourable result in such cases, I am not prepared to say, but think it highly probably it might do so. I feel quite confident, however, that in cases which are amenable to treatment, this will be found one of the most speedy and certain remedies. Of all the complaints for which mesmerism has been lauded as beneficial, there are none so conspicuous as epilepsy, as has already been referred to in the introduction. As the effects of hypnotism are so nearly allied to mesmerism, it would be superfluous for me to detail a number of cases, I shall therefore give only a few.

Case XLIX. A girl who had been liable to six or eight fits in 24 hours, had only one the day after she was first hypnotized, none for next five days, and was shortly quite well.

Case L. John Barker, aged 19 years, applied to me in August, 1842, for epileptic fits. He had first been seized with them when four or five years of age, at first every week or fortnight, but as he got older, became more frequent, so that, for some months previous to applying to me, he had had as many as three fits a week—had been under treatment at a public institution for two months before calling on me, and had a great variety of treatment, but derived no benefit, and was then told by the attendant, that he must never expect to get rid of them. He was subjected to my usual hypnotic operation for such cases, was operated on ten times altogether, and has had but one fit since

U

he was operated on; and that was the day after first operation. He had no medicine from me excepting three aperient powders. He has now been free of the fits for upwards of nine months.

Case LI. Mrs B., the mother of a family, had been subject to epilepsy for seven years, and notwithstanding every variety of treatment, allopathetic and homœopathetic, she had an attack at least once a month. From the time she was hypnotized she had no fit for four months, and has had none since.

Case LII. Miss B. had been subject to fits for nearly two years, latterly had as many as five and six a day; consulted me the end of December, 1842; was hypnotized seven times, and had no return of the fits for four months, when she had one, and in two weeks after a second.[1]

Hypnotism may be applied with great success in the treatment of various distortions, arising from weakness of certain muscles, or inordinate power or contraction of their antagonists; and I feel convinced, that by this means, we may rectify many of those cases which have hitherto been treated by section of the tendons or muscles. The success which I have already had, by this means, of treating *lateral curvature* of the spine, warrants me to speak very confidently on the subject, in most cases. *I feel convinced, there are very few recent cases which may not be speedily cured by hypnotism, without either pain or inconvenience to the patient.* Patience and perseverance will of course be necessary where the disease has been of long standing, and though in such cases the cure may not be perfect, the patient may be greatly improved by hypnotism.

[1] See Appendix II., Note 6.

The method of treating such cases is, first to induce the sleep, and then to call such muscles into action as are calculated to bring the body into the most natural position. By bringing these muscles into play during this condition, they acquire increased power, and ultimately are permanently strengthened. As one side of the chest is enlarged, and the other collapsed, I endeavour to restrain the enlarged side, by applying compression to it during the sleep, whilst the patient is directed to take deep inspirations, so as to expand the *opposite* side. I also endeavour to make the patient stand in a position *the very reverse of that which I consider to have been the chief cause of the curvature.* As already remarked, I feel convinced this method will prove very speedily successful, more decidedly so than any other mode of treatment I know of, and *especially in such cases as are accompanied with spinal irritation.*

Case LIII. The following is a case of its remarkable success with a young lady, 14 years of age, who had had the advice of some of the most eminent members of the profession in the provinces, and also in Dublin and London. She was first observed to become malformed when four years old. When brought to me on the 12th September, 1842, her chin rested on her breast, and there was no power of raising it, from the weakness of the recti muscles of the back, and contraction of the sterno-cleido-mastoid muscles. The dorsal part of the spine and shoulders was thrown backwards, the lumbar vertebræ and pelvis were thrown forwards, so that the deformity was very great, and the vigour of the mind, as well as of the body, was greatly impaired. She had no medicine nor external applica tion, but was hypnotized night and morning, and treated in the manner referred to, and the result was, that in six weeks she could hold herself so much better, that when the outline was taken, it was found that her spine was three inches nearer the perpendicular than when I first saw her. During this period, no mechanical means had been used, nor throughout any part of the time she was under my care were any resorted to, with the exception of a support for the chin, by way of remembrancer, till the habit of attention was acquired of supporting the head by mere muscular effort, which she now had the full power of doing. Nor should I omit to add, there was also a great improvement in the mental faculties.

Neuralgic pain in the heart and palpitation, I have also found to be relieved, or entirely cured, by neuro-hypnotism, more certainly and speedily, than by any other means. The following are examples :—

Case LIV. Miss Tomlinson, 16 years of age, I have already referred to. She had suffered severely from painful affection of the heart, with palpitation, which had resisted all treatment, and she had been prescribed for by eminent professional men, both physicians and surgeons. After being twice hypnotized, the affection of the heart disappeared, and has never returned but once, when it was immediately removed by hypnotism. It is now seventeen months since she was first operated on, and she is in perfect health.

Case LV. Miss Stowe, 22 years of age. I have already referred to her as one of the cases in which sight was remarkably improved by hypnotism. She had also suffered most severely from palpitation of the heart, accompanied with difficulty of breathing and dropsy, and various other symptoms which led the medical attendants, one of them an eminent physician, to pronounce the case hopeless, considering there was serious organic disease of the heart. After being twice hypnotized, all symptoms of affection of the heart disappeared (sufficient proof it had been only functional derangement), and she was speedily in the enjoyment of perfect health, and has been so now for the last twelve months, and that from hypnotism only. This patient had leucorrhœal discharge, which had resisted every remedy for years, and was so offensive as to cause suspicion she had malignant uterine disease. It was completely gone in a week, after being first hypnotized. She had no medicine excepting a single aperient pill occasionally. I should add, her hearing, as well as sight, was very much improved by it.

Case LVI. Mr —— had suffered severely from pain in the heart and palpitation. He was hypnotized with decided relief, and a second operation completely restored him, and he has kept well for the last eight months.

Case LVII. Miss —— had suffered much from palpitation of the heart, so that she could not ascend an easy stair without bringing on the most violent palpitation. I tested this before operating on her. After being operated on, caused her ascend the same flight of steps, which produced no palpitation, and she has never required the operation to be repeated.

Case LVIII. A young man had suffered much from valvular disease of the heart and palpitation and difficulty of breathing for four years, the consequence of a rheumatic fever. He could not walk more than twenty or thirty paces without being forced to stand or sit down. After being hypnotized for a short time he could manage to walk upwards of a mile at a stretch. In this case there was so much organic disease as precluded the hope of a perfect cure, but no means could have achieved for him what hypnotism did, and in such a short time too.

When considering the power of hypnotism in blunting morbid feeling, I may advert to its power of relieving, or entirely preventing, the pain incident to patients undergoing surgical operations. I am quite satisfied that hypnotism is capable of throwing a patient into that state in which he shall be entirely unconscious of the pain of a surgical operation, or of greatly moderating it, according to the time allowed and mode of management resorted to. Thus, I have myself extracted teeth from six patients under this influence without pain, and to some others with so little pain, that they did not know a tooth had been extracted ; and a professional friend, Mr Gardom, has operated in my way lately, and extracted a very firm tooth without the patient evincing any symptom of feeling pain during the operation ; and when roused, was quite unconscious of such an operation having been performed. He has extracted a second for this patient, and one for another, without their being conscious of the operation. To insure this, however, I consider that, in the majority of instances, it is quite necessary the patient should not, when he sits down, know or imagine the operation is to be performed *at that time*, otherwise the distraction of the mind, from this cause, may render it impossible for him to become hynotized deeply enough to render him *altogether insensible to pain*. The following case will illustrate this view.

Case LIX. Mr Walker called on me, stating he had been suffering from a violent toothache ; said he was anxious to have the tooth extracted, but that he suffered so much pain from the operation, on former occasions, that he could not make up his mind to submit to it, unless when hypnotized. He had been frequently hypnotized, and was highly susceptible of the influence. I told him I should be most happy to try, but that unless he could restrain his mind from dwelling *on the operation*, I might not be able to succeed in extracting the tooth, *entirely without pain*. He sat down, and speedily became hypnotized, but I could not produce *rigidity* of the extremities, nor *insensibility to pinching*, which in general were so readily induced in him. I therefore roused him, and told him the fact. He stated he went on as usual *to a certain point*, but then began to think, "now he will be putting the instrument in my mouth," after which the hypnotic effects went no farther. The pain was gone, and he left. In the evening he again called on me, when I tried him once more with the same results. I now aroused him, told him it could not be done with him reduced to a state of *total insensibility*, and that I should therefore extract it now that he was awake. I now extracted the tooth. He was conscious of my laying hold of it, but had felt so little pain that he could not believe the tooth had been extracted. Nor would he believe it till he had the tooth put into his hand. I now requested him to be hypnotized once more, when he became *highly rigid and insensible, in a shorter time than I had ever seen in him before*. From this, and other cases, I infer, that if it is intended to perform a surgical operation *entirely*

without pain, whilst in the hypnotic condition, the
patient's consent should be obtained for it to be done
sometime, but he ought on no account to know *when*
it is to be done, otherwise, in most cases, it would foil
the attempt.

However, that patients may be operated on with
greatly *less* pain even when in the *first* degree of
hypnotism, and whilst expecting an operation, is quite
certain, from the result of the case of Mrs ——, related
below, which I now refer to as Case LX. I have
also performed other operations under similar circumstances, and with similar results, namely, with *greatly
diminished* pain, although not *entirely without pain.*

Case LX. A lady had abscess connected with disease of the orbitar process of the frontal bone, had
the matter discharged by small puncture, the wound
closed by first intention and again opened, as required,
by the lancet. She experienced so much pain on
each occasion as to induce me to hypnotize her, after
which she made no complaint, although I durst not
carry it far owing to the state of the brain. On one
occasion I was anxious to ascertain how she would
feel by operating *without hypnotizing*, when the result
was so distressing, as to induce me always in future to
hypnotize her, before such operations, and then all
went on well.

Case LXI. An adult with worst variety of Talipes
varus, of both feet, had the first operated on in the
usual way, and the other whilst in the primary state
of hypnotism. The present ease and future advantage,
in respect to the latter operation, were most remarkable.
I have operated on upwards of three hundred club feet
now, and I am warranted in saying I never had so

satisfactory a result as in the one now referred to.

In cases of dyspepsia it is of the greatest service. Most patients feel the appetite greatly increased by being hypnotized, and that the digestion is more vigorous than before being operated on. All complaints, therefore, immediately connected with, or dependent on, indigestion, may be expected to be benefited by hypnotism. It is well known, many cutaneous diseases are of this class; and the following will illustrate the remarkable power exercised by hypnotism on this symptom, as well as several others associated with it :—

Case LXII. Mrs O., 33 years of age, the mother of a family, had been very nervous for fifteen years, with tremor of the arms, was easily alarmed, much disturbed by distressing dreams, and required being aroused several times every night from severe attacks of nightmare. She had also suffered severely from an inveterate eczema of the chest and mamma, and integuments of the abdomen, which, for five months, had resisted every remedy, both external and internal, under highly respectable medical men. The fingers of one hand were also affected with impetigo. She consulted me 31st August, 1842, when she was hypnotized, and was aroused greatly relieved from the distressing feelings of the head, and general nervousness. Her husband assured me, that on walking out with her same evening, had he not seen her, he could not have believed it was his wife who had hold of his arm, so much was the tremor of her arm improved, and she slept soundly all night without being troubled either by dreams or nightmare. She was hypnotized daily, and in a few days she was quite well, both as regarded her general health, and the obstinate skin disease ; and as she had no medicine nor external application, there could be no disputing that it resulted entirely from the influence of hypnotism.   She has been well nearly ten months.

Case LXIII. J. C., aged 40, had been severely afflicted for eighteen months with impetigo sparsa, extending from a little below the knee to near the toes.   He had also severe pain in the ankle joint, so that he had been disabled for work for eighteen months.   I hypnotized him, when he could walk better, after first operation, without his stick, than he could do immediately before with it.   In a few days the improvement was very remarkable, and within a week the disease of the skin was nearly well, and very little pain in the joint.   He was hypnotized almost daily till the end of the month, had no medicine, and no dressing but a little spermaceti ointment to prevent the cloth surrounding his leg from adhering to the sore ; and the skin disease being now quite well, and very little pain in the ankle joint, in a few days after he was enabled to resume his work.   He had undergone much treatment, under both public and private practitioners, but was becoming worse instead of better.   The immediate improvement in the appearance of the cutaneous disease, as well as feelings of the patients in the two last cases, were too obvious to admit of a doubt as to the remarkable powers of hypnotism.

The next cases I shall refer to, are those of permanent contraction or tonic spasm.   The following are interesting examples of this form of disorder, and the success of hypnotism in the treatment of them.

Case LXIV. Mr J. O., 21 years of age, called on me

1st October, 1842, complaining of a pain in the left
temple, a continual noise in the left ear, with occasional
shoots of pain, and the hearing of that ear very imperfect. He complained also of inability to open the
mouth so as to enable him to take his food comfortably, and that mastication caused great pain, so that
he frequently felt compelled to decline taking his
meals. Complaints had been coming on since previous Easter, and were becoming worse, notwithstanding he had been under the care of two medical men
up to the day before he called on me. That day, 10th
October, 1842, he could not eat breakfast but with great
difficulty, and had been compelled to take rice and milk
for dinner for two days, as he could eat nothing solid.
I found he could not permit the mouth to be opened
more than half an inch without great pain and difficulty; and besides the dulness of hearing already referred to, I found he had also very imperfect sight of
the left eye, which I tested very accurately. He had
not been aware of this until I called his attention to it.
I hypnotized him for about eight minutes, during which
I was enabled to open the mouth till the front teeth
were nearly two inches asunder, and he experienced
no inconvenience from me doing so. On being aroused,
all the pain in the temple was gone ; he could himself
separate the teeth one inch and three quarters, as accurately measured, in presence of four very intelligent
gentlemen who had been present during the operation ;
and he could move the jaws with the most perfect freedom, and without pain. The hearing was also much
better, and the sight of the left eye also most remarkably improved. 2nd October, called on me again, when
he stated he had been enabled to eat a good supper

after he left me the night before, and to take his breakfast and dinner with perfect comfort to himself; that
his hearing was much better; and the sight of both
eyes as nearly as possible equal. He had had no pain
in the temple since he was operated on, unless when
the mouth was opened to the utmost extent, and even
then it was trifling. He could now open the mouth
to nearly one inch and three quarters, *before* being
operated on to-day, and *after* the operation, to the
extent of two inches, that is, the front teeth were two
inches apart.

This patient called on me a few days after to be
operated on a third time, and retained the improvement noted above. He was to call again if he had
any relapse, but as he has not done so, I conclude
he continues quite well, and it is now nearly seven
months since I last saw him.

I shall now refer to other cases of spasmodic affection, which are most interesting, as they afford us
strong grounds to hope that Tetanus, Hydrophobia,
and other analogous affections, may be arrested and
cured by this agency.

Case LXV. A girl was seized with violent tonic
spasm of the right hand and arm, and side of the face.
A respectable surgeon was consulted, who ordered a
blister to the nape of the neck, medicine, fomentation,
and liniments to the parts affected. The symptoms
became more urgent, and they sent for the surgeon
again, but as he was out, and as they were much
alarmed, I was consulted. The blister had been applied, but the medicines had not been used as directed.
I found the hand so firmly clenched that it was impossible to open it, the arm so rigid it could not be

moved ; but, knowing the efficacy of my new remedy, I hypnotized her, and in *two minutes*, with the most perfect ease, I unlocked the hand, and removed the other spasmodic contractions, and she was instantly quite well, and has continued so ever since, now more than a year.

I shall only record one additional case, and a more remarkable or satisfactory one I think could scarcely be adduced. I give the case as correctly recorded by the patient's father, in a letter he sent for my approval, previous to having it sent to be recorded in some periodical. I preferred having its publication postponed, and now give it precisely in his own words.

Case LXVI. Miss Collins of Newark. " My daughter, 16 years of age, had been afflicted for six months with a rigid contraction of the muscles on the left side of the neck, to so great a degree, that it would have been impossible to insert an ordinary card between the ear and shoulder, so close was their contact; and consequently she was rapidly becoming malformed. She had had the best advice to be procured in the country, and I had taken her to London with a written statement of the treatment previously employed, and had the opinion of Sir Benjamin Brodie, who approved of what had been done, but gave no hope of speedy relief.

" In consequence of seeing a report of a lecture given on the subject by Mr Braid, surgeon, St Peter's Square, Manchester, and a letter written to that gentleman by Mr Mayo of London, I went with her, by the advice of Dr Chawner, who indeed accompanied us, and placed her under the care of Mr Braid on Thursday evening, the 24th March last (1842). In less than a minute after that gentleman began to fix her attention, she

was in a mesmeric (neuro-hypnotic) slumber, and in
another minute was partially cataleptic. Mr Braid
then, without awaking her, and consequently without
giving her any pain, placed her head upright, which I
firmly believe could not, by any possibility, have been
done five minutes before, without disruption of the
muscles, or the infliction of some serious injury, and I
am thankful to say, it not only continues straight, but
she has the perfect control over the muscles of the
neck. A nervous motion of the head, to which she had
been subject after her return from Manchester, has
entirely ceased, and she is at present in excellent health.
It is necessary to remark, that at Dr Chawner's recommendation she was frequently watched while asleep,
but not the slightest relaxation was observed in the
contracted muscles.

"Many respectable persons can bear testimony to
the statements herein made.

(Signed)   JAMES COLLINS."

"NEWARK, 11th May, 1842."

I have been informed that some very absurd reports
have been circulated, even in the metropolis, as to my
mode of operating on this patient, namely, that I had
exhibited a vast display of gesticulations and hocus
pocus, in order to work upon her imagination. SUCH
STATEMENTS ARE UTTERLY UNTRUE. I simply desired
her to maintain a steady gaze at my lancet case, held
above her eyes in the manner pointed out at page 109
of this work, and after the eyes had been closed, and
the limbs extended for about two minutes, I placed my
left hand on the right side of her neck, and my right
hand on the left side of her head, and, by gentle means,

gave a new direction to the sensorial and muscular power, and was thus enabled by *art*, rather than mechanical force, in less than half a minute, to incline the head from the left to the right of the mesial plane. The muscular contraction being thus excited on the right side of the neck, in muscles which had been inactive for six months previously, was the surest and most natural mode of withdrawing the power from their antagonists, and reducing the spasm of the contracted muscles on the left side. After allowing the patient to remain two minutes supporting her head, now inclined towards the right, by her own muscular efforts, to give them power on the principle already explained, I aroused her in my usual way, by a clap of my hands. The patient's father, and Dr Chawner of Newark, were present the whole time, and to them I appeal as to the correctness of this statement, and in refutation of the vile, unfounded calumny above referred to.

After the lapse of a year Mr Collins was so kind as write, to inform me his daughter continued in perfect health, with complete control over the muscles of the neck.

I could easily adduce many more interesting cases, but trust those already recorded may be sufficient to prove that hypnotism is an important addition to our curative means, and a power well worthy the attentive consideration of every enlightened and unprejudiced medical man.

APPENDIX I

*Notes by the Editor, chiefly derived from the Later
Writings of James Braid.*

NOTE I (p. 74).

THE analogy here instituted between the effects produced by hypnotism and those of opium and other
anæsthetics, is further developed in the treatise on
"Trance, or Hybernation." Though the facts, at the
present day, are a matter of common knowledge, the
additional observations of Braid, at a period when they
had an aspect of novelty, are not without interest,
and will therefore bear quotation. "The phenomena
realised by the use of ether and chloroform, in like
manner, manifest similar states of anæsthesia more or
less profound, according to the quantity exhibited and
constitution of the patient, and have given a ready
credence to phenomena realised by these means which
had been by many pronounced cases of rank imposi-
tion when induced by mesmeric or hypnotic methods.
The peculiar conditions, also, induced by the use of
Bangue, Hachisch, and Dawamese, in the East, all tend
to illustrate certain conditions of the nervous system
producible by artificial contrivances. Thus, a slight
dose of these produces mental hallucination, with some
degree of control over the train of thought—a sort of

half-waking dream. As the effects advance, the imagination becomes more and more vivid, and a rapid succession of ideas passes through the mind, assuming all the force of present realities." Quoting Dr Carpenter, Braid proceeds to describe this torrent of disconnected ideas, which increases till the attention is completely arrested and the mind wholly withdrawn from the contemplation of external realities into the consciousness of its own internal workings. This consciousness is always far greater than that which exists in ordinary dreaming. It is the state of conscious dreaming. " The succession of ideas has at first less of incoherence than in ordinary dreaming, the ideal events not departing so far from possible realities ; and the disorder of mind is at first manifested in errors of sense, in false convictions, or in the predominance of one or more extravagant ideas. These ideas and convictions are generally not altogether of an imaginary character, but are called up by external impressions, which are erroneously interpreted by the perceptive faculties. The error of perception is remarkably shown in regard to time and space; minutes seem hours, hours are prolonged into years, and at last all idea of time seems obliterated, and past and present are confounded together as in ordinary dreaming; in like manner streets appear of an interminable length, and the people at the other end seem to be at a vast distance ; still there is a certain consciousness of the deceptive nature of these illusions, which, if the dose be moderate, is never entirely lost." Some stages of natural dreaming are occasionally characterised, however, by this consciousness of illusion. Braid goes on to particularise the effects of the full dose, when the

mind is completely withdrawn from the distinct comprehension of the world without. "The power of the
will over the current of thought is in like manner
suspended," the contrast with ordinary dreaming being
chiefly in the stronger expression of the feelings,
which still "take their tone almost entirely from external impressions." From this inhibition of the attention, the patient, under the influence of a stronger
dose, will pass into a state of "profound narcotism,"
suitable to the performance of painless surgical operations, precisely as under chloroform, while in India,
Braid is careful to add, "the experience of Dr Esdaile
has proved mesmerism equally potent for like purposes."
He remains, however, of the opinion that chloroform
is the more certain and speedy agent for effecting
painless operations in our own country. While affirm-
ing that the phenomena which he has recited can be
produced in as genuine and true a manner by hypnotic
or mesmeric processes as by the ingestion of the medicinal agents referred to, he explains the prevailing
scepticism on the subject of such phenomena by the
extravagance of mesmerists, "who have contended for
the reality of clairvoyance in some of their patients,
such as seeing through opaque bodies, and investing
them with gifts and graces of omniscience, omni-
presence, mesmeric intuition, and universal knowledge,
—pretensions alike a mockery of the human understanding, as they are opposed to all the known laws of
physical science." The lapse of nearly half a century
may not have placed clairvoyance among the established facts which are universally recognised by science,
but as it was obvious then that Braid unreasonably
exaggerated the powers attributed to the faculty for

the purpose of denying its existence, so there are few
now who would reject its possibility on the ground
that it was in conflict with natural law.   The scientific
verdict of the moment is that of " insufficiently established," and by no means that of "impossible."

NOTE 2 (p. 81).

The opportunity does not appear to have been
afforded and there is no record in the later annals of
hypnotism or mesmerism to give any colour to the
pretensions that either lock-jaw or rabies can be cured
or relieved by resort to these agents.

NOTE 3 (p. 86).

Nine years later this point of view is reinforced in
the pamphlet entitled " Magic, Witchcraft, and Animal
Magnetism."   Replying to the strictures of J. C. Colquhoun, Braid says : " Whilst I have had the most satisfactory proof of the value of the hypnotic mode of
treating some forms of disease, when it is judiciously
applied, I repudiate the notion of holding it up as a
panacæa or universal remedy ; or believing that the
efficacy of medicines mainly depends upon ' the peculiar
idiosyncrasy, or magnetic temperament of the pre-
scriber ' "—an opinion registered by Colquhoun.   Braid
adds : " Whilst I most readily admit that the efficacy
of medical treatment may be greatly aided by the
peculiar manners, looks, and language of the person
who prescribes the medicines, and the confidence engendered by these means in favour of his prescriptions,
still, I feel assured that there is *a positive and obvious
effect to be expected from some medicines, altogether irrespective of the physical or mental qualities and manners*

of the individual who prescribes them." This recalls the proverb of " measures not men " with the qualification of Bailey's " Festus " :—" Something at times depends upon the man." The time and experience to which Braid appeals in the text have accorded hypnotism a distinctive therapeutic sphere which, however, by the narrowness of its range, has so far justified the prudent reticence of its discoverer.

NOTE 4 (p. 91).

The idea of criminal suggestions acting post-hypnotically was unknown, as it need not be said, in the days when Braid recorded his experiments, and the reference in the text is confined to the obvious opportunity which the passive state of a subject may offer to an unscrupulous operator. There seems to be nothing in the considerations advanced which does not apply also to the mesmeric state. Mr James Coates, in his recent work which has been the subject of a previous reference, using Hypnotism and Animal Magnetism as interchangeable terms, and following the lead of De Courmelles in his rejection of the supposed lessons of laboratory crimes, dismisses also the stories of fascination and seduction on the authority of his personal observation that all subjects manifest in the hypnotic state a keener sense of right and wrong than they exhibit in ordinary life. " The latent moral powers and the will to do right in each subject may be calculated upon in the, apparently, most hopeless cases." On the whole, there are no dangers in the one condition from which immunity may be affirmed in the other, nor can it be said that one offers more than the other a special protection by a superior capacity for awaking the higher

moral faculties of the subject. There are phases in both
which seem obviously liable to abuse, but the known
cases of abuse are at the same time exceedingly few.
As to the larger possibility, it seems to exist only in the
fields of romance and in the dramatic entertainments
of the laboratory. Dr Charcot stated in 1890, that
" there is not a single example of such a crime," and
the position remains unchanged at the present moment.

NOTE 5 (p. 93).

With this opinion the majority of medical men continue to agree, while non-professional operators dissent
now as they dissented in the days of Braid. Something
must be allowed in both cases for the colouring of the
interests which they represent. There are conceivably
many lay practitioners who may be as qualified as
many medical men, and there are some who might be
more qualified than some doctors. Mr Coates, representing the able " lay practitioner," would appeal to his
record of successes, and appeals of this kind cannot be
lightly set aside. " Fitness and experience," he observes, " not medical registration, can alone qualify a
person to hypnotise." But there will be no opportunity
for the experience of lay operators if their experiments
are pronounced illegal. We must beware of creating
monopolies which may disqualify competent persons
on arbitrary grounds ; the hostility to the layman is
largely based on possibilities which are admitted only
by the minority of alarmists. At the same time, it must
be acknowledged that the considerations offered by the
layman would warrant him quite as much in the unre-
gistered practice of surgery. It is at least obvious that
the gifted hypnotist might do worse than take his degree.

NOTE 6 (p. 94).

To this definition which is substantially repeated in the preface to " Observations on Trance," Braid there adds the following qualification, which is useful as a short summary of his views :—" I do not allege that this condition is induced through the transmission of a magnetic or occult influence from my body into that of my patients ; nor do I profess, by my processes, to produce the higher phenomena of the Mesmerists. My pretensions are of a much more humble character, and are all consistent with generally admitted principles in physiological and psychological science. Hypnotism might therefore not inaptly be designated, *Rational Mesmerism*, in contra-distinction to the *Transcendental Mesmerism* of the Mesmerists." The last words exhibit an illiberal distinction, and seem also to convey much more than was intended by their writer, who does not anywhere state that he has abandoned the reasonable position adopted, as regards the higher phenomena of mesmerism, in the first chapter of the present work.

NOTE 7 (p. 126).

"The *hypnotic sleep*," says Braid, in his pamphlet on "Magic, Witchcraft," etc., "is the very antithesis or opposite mental and physical condition to that which precedes and accompanies *common sleep ;* for the latter arises from a diffusive state of mind, or complete loss of power of fixing the attention, with suspension of voluntary power. . . . The state of mental concentration, however, which is the basis of the *hypnotic* sleep, enables the subject to exhibit

various passive or active manifestations, such as insensibility or exalted sensibility, rigidity or agility,
and entire prostration or inordinate energy of physical
power, according to the trains of ideas and motives
which may arise spontaneously in his mind, or be
addressed to it by others, through impressions on his
physical organs."

Again, in "Hynotic Therapeutics," Braid distinguishes the principal difference between the hypnotic,
or nervous, and common sleep. " In passing into common sleep the mind is diffusive or passive, flitting
from one idea to another indifferently, thereby rendering the subject unable to fix his attention effectively
on any regular train of thought, or to perform any
acts requiring much effort of the will. The consequence is this, that a state of passiveness is manifested during the sleep, so that audible suggestions
and sensible impressions addressed to the sleeper, if
not intense enough to awake him entirely, seldom
do more than excite a dream, in which ideas pass
through his mind without exciting definite physical acts; but, on the other hand, the active and
concentrated state of mind engendered by the processes for producing the *nervous* sleep are carried *into*
the sleep, and, in many instances, excite the sleeper,
without awaking, to speak or exhibit physical manifestations of the suggestions received through words
audibly uttered in his hearing, or ideas previously
existing in his mind, or excited by sensible impressions made by touches or passes of the operator,
which direct the attention of the sleeper *to* different
parts, or excite into action certain combinations of
muscles, and thereby direct his current of thought."

NOTE 8 (p. 129).

These methods of dehypnotising were afterwards discriminated according to the intention of the operator. Braid came to learn that much depended upon the mode of arousing the patient, in respect of the results which were looked for. "If I wish," he says in "Hypnotic Therapeutics," published in 1853, "any predominant idea, or any physical change which has been induced, to be carried strongly into the waking condition, I arouse the patient abruptly, by a clap of my hands near to his ear, when he is in the full height of the desired condition; but if tranquillising is the object in view, then he had better be aroused slowly and softly, such as by gently wafting the face of the patient with the hand, or with a fan, or an open handkerchief; or by placing the balls of the thumbs gently on the eyes of the patient, or on his eyebrows, and then carrying them laterally a few times so as to produce gentle friction, to which may be added gentle fanning when required."

NOTE 9 (p. 156).

In his pamphlet on "The Power of the Mind over the Body," the author reaffirms this position in the following closing words:—"The results of my experiments satisfactorily prove the *unity* of the *mind;* and the remarkable power of the soul over the body."

NOTE 10 (p. 165).

"As regards sympathy and imitation," says the pamphlet on Magic and Witchcraft, "every medical man is well aware of their wonderful power, especially on certain temperaments. From the wag who sets a whole company yawning, against which the very dog below his master's table cannot protect himself, to the recent public exhibition of experiments in 'electro-biology,' we have abundant proofs of their power, as well as of the influence of audible suggestions. In particular, we have the history of its extensive prevalence in the middle ages, in the dancing mania, the St Vitus's and St John's dances. Again, the dancers who had the disease induced by the alleged poisonous bite of the Tarantula, or venomous spider, likewise spread extensively in Italy, 'where, during some centuries, it prevailed as a great epidemic.' . . . Even in recent times, and in our own country, the power of sympathy and imitation was vividly pourtrayed in the south and west of England, where, from this influence, within a short period, about 4000 people became affected by a violent convulsive malady. It is recorded of this epidemic that 'hundreds of people who had come thither, either attracted by curiosity, or a desire, from other motives, to see the sufferers, fell into the same state.' The like condition was also manifested very lately in the Shetland Isles. . . . Let those who deride the notion of *their* being susceptible of such impressions, from *such* causes, have a care that they do not trifle too confidently with the attempt to prove how far *they* can surrender their self-control, for a time, for the purpose of imitating others. They may very probably, in the end, find to their cost and sorrow, that the *will* having been for a time surrendered into another's keeping, may not return at their bidding ; and that what they believed to be a mere delusion or fraud in others, has proved to be a grave reality with themselves. They ought to remember the fate of the distinguished prelate, Jo. Baptist Quinzato, Bishop of Foligno, who 'having allowed himself, by way of a joke, to be bitten by a tarantula, could obtain a cure in no other way than by being, through the influence of the tarantula, compelled to dance.' Now, it is fully ascertained that the bite of the tarantula is not at all poisonous, and the bishop was fully persuaded of this when he perpetrated the joke ; still, the idea having at length taken possession of his mind, that it possibly *might* be poisonous after all, he became a victim to this *imaginary* poison.   From the state of public opinion in that age, it has, moreover, been recorded, 'that even the most decided sceptics, incapable of guarding themselves against the recollection of *what had been presented to the eye*, were subdued by a poison, the powers of which had been ridiculed, and which *was* IN ITSELF INERT *in its effects.*' . . . The beneficial effects of dancing, for dissipating the baneful influence of fixed morbid impressions and associations, is proved by the experience which has been had on this point, since this mode of amusement and exercise has been introduced into some of our public lunatic asylums. The effects of sympathy and imitation and excessive excitement, manifested amongst certain religious sects at their camp meetings in North America, in recent times, prove that, with all our boasted advancement in civilisation, beyond those who inhabited Europe during the middle ages, the human mind is still susceptible of these strange mental delusions and physical manifestations, excited entirely through the power of sympathy and imitation in those who place themselves in circumstances favourable for their development."

NOTE 11 (p. 189).

"I have frequently hypnotised *insane* patients, and with marked advantage; but hitherto I never could succeed in inducing sleep *in a decidedly fatuous case.* This I attribute to the great difficulty of arresting their attention; and after numerous and prolonged trials, being unwilling to waste more time, I have abandoned them, although it is highly probable that a continuance of the means might have ultimately succeeded."—Braid's "Observations on Mesmeric and Hypnotic Phenomena," in *The Medical Times*, April 13, 1844.

NOTE 12 (p. 215).

"At p. 148 of my work on hypnotism, I stated that the phrenologists' point 'eventuality' I had strong grounds for believing to be the chief seat of memory. I was led into this error from finding that patients, when touched on that point, could always answer correctly any question about past events, although unable to do so *without* such contact. The circumstances now noted (*i.e.*, the influence of contact, during hypnotism, in exciting memory), prove I was so far in error, as *contact with any part of the body* would have produced similar results."—"Observations on Mesmeric and Hypnotic Phenomena," *Medical Times*, April 13, 1844.

NOTE 13 (p. 219).

"I feel pretty confident that whoever will undertake the investigation of hypnotic phenomena with a candid mind, and untrammelled by any previous prejudices in favour of the mystical and transcendental, may very soon satisfy himself that the real origin and essence of the hypnotic condition is the induction of a habit of abstraction or mental concentration, in which, as in reverie or spontaneous abstraction, the powers of the mind are so much engrossed with a single idea or train of thought, as, for the nonce, to render the individual unconscious of, or indifferently conscious to, all other ideas, impressions, or trains of thought."—*Magic, Witchcraft*, &c., pp. 53, 54.

In the pamphlet on " Hypnotic Therapeutics," Braid describes, perhaps more clearly than elsewhere, the consequence of this mental concentration ; it " intensifies, in a correspondingly greater degree, whatever influence the mind of the individual can produce upon his physical functions during the waking condition, when his attention is so much more diffused and distracted by other impressions. Moreover, inasmuch as words spoken, or other sensible impressions made on the body of an individual by a second party, act as suggestions of thought and action to the person impressed, so as to draw and fix his attention to one part or function of his body, and withdraw it from others, whatever influence such suggestions and impressions are capable of producing during the ordinary waking condition, should naturally be expected to act with correspondingly greater effect during this *nervous sleep*, when the attention is so much more concentrated, and the imagination and faith, or expectant ideas in the mind of the patient, are so much more intense than in the ordinary waking condition."

Two cases cited by Braid seem to be of especial interest in this connection.

" Having told a gentleman that the expectant idea

in the mind of a patient was quite adequate to produce
a corresponding change in the physical function of any
organ or part of the body to which it was directed, he
expressed his incredulity. I asked him if his wife was
not then nursing, to which he replied she was ; and I
therefore offered to prove my position, if he chose, by
causing an increased flow of milk to come into one of
her breasts, by directing her attention particularly to
that breast during the sleep. This gentleman's wife
had been a patient of mine some eight months previously, and was then cured of violent headaches by
hypnotism ; and I knew she was one of those subjects
who pass into the second-conscious or full state, and
upon whom the power of suggestion manifests its
greatest influence. The lady was sent for, and asked
if she had any objection to be hypnotised, for her
husband to have an opportunity of seeing her in that
state. She readily gave her assent, and whilst standing
on her feet, I held my lancet case over her head,
in my usual way, and requested her to gaze upon
it, and speedily her eyelids closed, with the twitter
peculiar to the hypnotic sleep. After she had remained in this state a little while, I gently drew the
tips of my fingers two or three times over the left
mamma, when the patient slowly raised her left arm
towards her breast. I then enquired, What is it? To
which she replied, ' Baby.' What about Baby ? To
which she answered, ' Oh, this is so tight,' pointing to
her left breast. In this state I allowed her to remain for
a few minutes, her mind riveted to the idea of her baby,
and the fulness of her breast. With a clap of my hands
I now aroused the patient, who had no recollection whatever of anything said or done when she was asleep. I

asked if any part of her body felt different from its usual condition? To which she replied, pointing to the left breast, this breast feels very tight. I asked her what had made it so? To this she replied, she could not tell, but that it felt so. Her husband now remarked, ' That is what Mr Braid said he would do—he said he could bring a rush of milk into it.' To this the lady replied, ' That will be no easy matter, for my baby is fourteen months old, and I have scarcely any milk.' I requested her to bring baby and try, as I felt assured that *now* there would be no lack of milk in that breast. The baby was applied to that breast, and, notwithstanding he was fourteen months old, the flow of milk was so copious that it nearly choked him. A few days thereafter this lady complained that I had disfigured her, as I had made her over-protuberant on the left side. I said I can soon settle that matter, for, by putting you to sleep again, I can take it down as readily as it was increased in size during former sleep. She most willingly assented to this, but when she was asleep, instead of taking it down (which a suggested idea to that effect would have done), I acted on the other breast in precisely the same manner as on this left breast, and with precisely similar results. The most important point, however, still remains to be told,—viz., that although her child was fourteen months old, and before being hypnotised she complained of having had very little milk, these hypnotic processes had given such a stimulus to the mammæ, that this lady was enabled to continue to suckle her child from an overflowing breast for *six months longer*.

"Another important fact I have to communicate connected with this case. This patient was one of those

ladies who menstruate during lactation, and, previous
to my hypnotising her the first day, she had gone two
weeks beyond the usual period for the appearance of
the catamenia. The stimulus to the mammæ, however,
through sympathy, had brought on that discharge also
within half an hour after she was hypnotised."

It will be observed that this case illustrates the statement by Braid, made in "Observations on Trance" and
elsewhere, as to the influence of hypnotism upon the
catamenial function. See "Biographical Introduction,"
p. 47.

The second case illustrating the influence of mental
impression on physical action was very highly regarded
by Braid himself, though the chain of causation does
not seem quite so clear. "The patient was one of those
subjects who pass into the second-conscious state of
hypnotism, and had been cured by hypnotism of paralysis, loss of sense and motion, of one side of the head
and face. The following effect of the expectant idea,
however, relates to what occurred when she was in
the waking condition. This patient, Mrs ——, was
the mother of three living children, the last of which
was a cross birth, delivery being accomplished with
great difficulty. The two subsequent births were of
largely developed children, both still-born, both having
been shoulder presentations, the labour far advanced,
and the shoulder and arm advanced within the pelvis
before medical assistance arrived. Upon careful examination of the bones of the pelvis of this patient, it
was clearly ascertained that there was such advancement forward, and depression of the promontory of the
sacrum and lumbar vertebræ, as to preclude the hope
of her ever giving birth to a full-sized living child;

and, therefore, when she again became pregnant, I explained how matters stood to her husband, as well as to the patient, and recommended that premature labour should be induced, as affording the only chance of her bearing another living child, and as affording the greatest safety, moreover, for the mother. Both parties were perfectly satisfied to abide by my decision on this point, so that I was to consider myself at perfect liberty to act in the matter as I thought best, both as to the method to be adopted for accomplishing such purpose, and also in regard to the time when I was to induce premature labour. About two weeks beyond the seventh month was the period which I had fixed on for inducing labour. I had seen the patient a few days before this period, and found her in excellent health, experiencing no inconvenience of any sort. I told her that in three or four days I intended to do something for her to bring on labour, as had previously been agreed upon should be done. She was quite agreeable to this proposal, and seemed to entertain no anxiety whatever on the subject. In two days thereafter, however, I was sent for to the patient, and ascertained that the mere mental impression had been sufficient to bring on labour, for the *os uteri* was not only fully dilated, but, as in three former labours, the shoulder was presenting. In this case, from the small size of the infant, I was enabled with great ease to turn and deliver the mother of a living child."

NOTE 14 (p. 287).

"Many sceptics have endeavoured to throw discredit on the importance of hypnotic processes as a means of cure, by attributing the whole results to the

power of *imagination*. They are willing to admit that
certain effects are produced, and are content if they
can only damage the importance of the facts, by
associating them with what *they* consider a *bad name*.
Their admission of the *facts*, however, is something;
and we shall now devote a few moments to the consideration of the power of imagination over the physical
organism of man. Those who suppose that the power
of imagination is merely a mental emotion, which may
vary to any extent without corresponding changes in
the physical functions, labour under a mighty mistake.
It is notorious to those who have carefully studied this
curious subject, that imagination can either kill or
cure; that many tricks have been played upon healthy
persons, by several friends conspiring, in succession, to
express themselves as surprised, or sorry, or shocked
to see them looking so ill; and that very soon a visible
change has come over the patients, and they have
actually gone home and been confined there for days
from bodily illness thus induced. Not only so, but
there are even cases recorded, in which we have the
best authority for the fact, where patients who were
previously in perfect health, have actually died from
the power of imagination, excited entirely through the
suggestions of others. Nor are the suggestions by
others of the ideas of health, vigour, hope, and improved looks less influential with many people for
restoring health and energy both of mind and body.
Having such a mighty power to work with, then, the
great desideratum has been to devise the best means
for regulating and controlling it, so as to render it
subservient to our will for relieving and curing diseases.
The modes devised, both by mesmerists and hypnotists,

for these ends, I consider to be a real, solid, and important addition to practical therapeutics; and not the less curious and important that it is done simply through appeals to the *immortal soul*, to assert and demonstrate its superiority and control over the *mortal body*."

APPENDIX II

Containing Notes of further Cures by Hypnotism from the Later Writings of James Braid.

NOTE I (p. 260).

In "Observations on Trance" there is an account of the removal of an olfactory hallucination which it is interesting to connect with the restoration of the sense of smell by the same agency. "In July, 1849, when visiting a child, the mother of whom was forty years of age, and had had nine healthy children, the mother stated that she herself had been tormented for four days with the most disgusting smell. The impression had arisen from her going into the house of a relative whose body had been kept till it was in a state of advanced decomposition. From the moment the impression was made on her olfactory nerves on entering the house of the deceased, she had never been able to get rid of it, whether within doors, or in the open air, notwithstanding she had resorted to the use of all sorts of fragrant scents, to sal volatile, to snuff, and even to the fumes of tobacco and burning lucifer-matches, and every other contrivance she could think of. Still all had proved in vain, the disgusting odour being still as vivid as the first moment of its impression. The patient, therefore, was glad to try any method I might suggest, and consequently I threw her into the hypnotic state, during which I kept her mind constantly dwelling on the idea of fragrant odours, by exciting the idea in her mind through auricular and muscular suggestion, according to the principles laid down in my writings on hypnotism. I took care to awake her with the mind actively engaged with these agreeable ideas, and the result was, that the moment she was aroused from the sleep, she exclaimed with delight, that she *now* enjoyed a smell *as delightful* as the former had been disgusting. From that moment this patient was never again annoyed with the disgusting odour, not even when, on several occasions, she attempted, by way of experiment, to recall it."

NOTE 2 (p. 269).

By far the most remarkable cure of spinal distortion is that described by Braid in a note to " Observations on Trance." The patients were sent him by Dr Garth Wilkinson, who had previously made his acquaintance to ascertain at first hand his views on hypnotism, and had been himself operated on with success. They were two girls aged respectively sixteen and fourteen years, who had been under treatment for several years. The affliction was severe, and the treatment included reclining night and day for sixteen months, the counter extension of pulleys and weights attached to the body, the use of the procrustean bed, and numerous other appliances. Notwithstanding all these efforts the deformity increased, and there was complicated distortion of both sides of the chest. At the first hypnotic operation both patients were rendered cataleptic, retaining, however, a perfect recollection of all that took place.

" Next morning they were again operated on, when they
were not only rendered cataleptic, but also talked and
wrote, and exhibited mental emotions through muscular
excitation, of all which they remembered nothing when
awake, but on being sent to sleep again they remembered every particular—thus manifesting double consciousness.   In the evening of that day I had the steel
apparatus removed before operating on my patients,
because I then intended to turn to account, in rectify-
ing the spinal distortion, the peculiar susceptibility
which had been induced by the previous operations.
The instant the apparatus was removed, the result of
the erroneous mode of treatment manifested itself"—
Dr Garth Wilkinson, it should be observed, had only
recently become their medical adviser.   "For there
was no muscular power to support the body in the
proper position, and the falling down and distortion
of both patients was most painful to witness."   In the
one case there was a contraction of the left side and a
corresponding projection of the right shoulder ; in the
other, a double curvature.   "Having hypnotised these
patients, and called a cataleptiform degree of energy
into the morbidly weakened muscles, and thus brought
the body into a more favourable position, it is impossible for me to describe the astonishment of the
mother and the ladies' maid. . . . Their surprise was
still greater, however, when they found that after the
patients were aroused they still retained an increased
degree of power, by which they were enabled to support
themselves. . . . Twice each day these operations were
repeated, and with such marked improvement, that
without the slightest inconvenience, in less than a week,
they could walk about from half an hour to three-

quarters without any support from stays. . . . After I had attended these patients for a week, the mother called my attention to the contracted state of her elder daughter's arms, both of which she had been unable to extend fully for upwards of four years. . . . I immediately acted on the right arm so as to ensure the cataleptiform state of the extensor muscles, and on arousing the patient a few minutes thereafter, it was ascertained that she could then extend the arm completely by her own voluntary effort. By then operating on the left arm an equally satisfactory result ensued." This account appears at the end of the work on Trance, and was added while passing through the press ; it was Braid's most recent case. Whether the cure was completed and permanent did not transpire.

NOTE 3 (p. 285).

The last pamphlet of James Braid, that on the " Treatment of Certain Forms of Paralysis," contains some additional cases of the cure of this disorder entirely by hypnotism. Among these the following is perhaps most deserving of remark.

" A young lady, aged 16, of Newark, Notts., was brought to me by Dr Chawner and her father. For twenty-six weeks her head had been drawn towards her left shoulder so closely that a card could not be introduced between them. All possible contrivances had been resorted to in vain, the cramp continuing even when sleep was accompanied by an opiate. A consultation with Sir Benjamin Brodie proved unsuccessful, nor could I move the head from the shoulder by any force which I exerted. I therefore hypnotised the sufferer, holding my lancet-case over the forehead at a

distance of ten or twelve inches, and causing her to fix
her eyes upon it. Presently her eyelids closed, the
symptom being accompanied by the winking custom-
ary to the induction of the hypnotic state. I then raised
her arms and legs to encourage rigidity, thus also
occasioning an agitation in the sensorium and spinal-
marrow. When after several minutes she had passed
into a condition which was favourable for operation on
the muscular system, I began to titillate the skin
softly on the right side of the throat, causing an activity
of the underlying muscles. This would, I knew, be the
shortest and safest way of removing the contraction of
the left side of the throat and the muscles of the neck.
I was now able, by art and not by force, to lift the head
and incline it towards the right shoulder, altering the
position at will and without difficulty. Leaving the
patient alone for a few minutes, I awoke her at length
by a slight blow behind the ear, the head still retaining
an erect position. As a slight crookedness of the
vertebral column remained, I hypnotised her subsequently when standing upright, and endeavoured during
the sleep to straighten the vertebral column by manipulation. This also was successful, and on again awaking
her, the head and vertebral column continued in their
natural position." Whether the cure was permanent
does not appear from the narrative.

NOTE 4 (p. 297).

Another successful treatment of an aggravated case
of rheumatism is recorded in " Observations on Trance.'
" On the 28th of March, 1843, I was requested by a
philanthropic gentleman (the individual here referred
to was the late Hon. and Rev. William Herbert, Dean

of Manchester) to extend my charitable sympathy to a poor woman of the name of Barber, and by the power of hypnotism to relieve her of a severe rheumatic affection, from which she had been suffering for several months. She was forty-four years of age, and a most pitiable object, suffering severely from pulmonary affection, as well as from rheumatism. With the latter affection she became afflicted about the beginning of winter; about the end of December, 1842, had been entirely confined to her bed for five weeks, after which she was able to get up; but the flexors of the legs and toes were so contracted that she could not extend them, and it was with great pain, as well as with diffi-culty, that she crawled about her apartment. Her hands, wrists, and fingers were also much affected; so that she was very helpless. Her pain was not only severe, but unremitting either by day or by night. After being hypnotised the first time, during which I endeavoured to regulate the irritated condition of the muscles, she was enabled to straighten her legs and toes, and move her wrist and fingers, could walk with greater freedom, and expressed herself almost entirely free from pain both of legs and arms before I left her house. After *five* operations, as is well known to many, she was so well as to be able either to walk or run across her room, and even to step on a chair, with either foot first, without assistance. I operated on her thirteen times altogether, and she remained, to my certain knowledge, free from rheumatism up to the end of December, 1843, and I have reason to believe had no other attack of it subsequently, as I heard nothing farther of her, and she had my assurance, as well as that of the late Dean, that I was willing to give her any

further assistance of the sort she might require at any
time, as an act of charity."

NOTE 5 (p. 303).

The cure of a persistent headache of several years'
standing is also given in the "Observations on Trance."
It was the case of a lady fifty-four years of age, and the
mother of a numerous family, who had been grievously
afflicted, practically without intermission, for the previous fourteen years. "She stated, moreover, that she
suffered from pain and smarting of her eyes, and that
her sight had become so affected, that she could not
then see to read a sign across the street. I recommended her to try the effect of hypnotism, which I
believed calculated to afford her relief for *both* affections.
To this she readily consented, and, to her great delight
and surprise, on being aroused from the first hypnotic
sleep, she found herself *quite free from headache, for the
first time for fourteen years*. Next day I found she had
a slight return of headache, and, for a few days thereafter, less and less of it daily, during which period I
hypnotised her daily, and, in the course of one week,
the headaches were entirely gone, and never returned
since, unless for a short time occasionally, as will occur
to any one ; and the pain and smarting which she
formerly felt in her eyes, also entirely disappeared, and
her sight became greatly improved, and has continued
so ever since." On a subsequent occasion she was
cured in the same manner of acute diarrhœa.

NOTE 6 (p. 306).

The following important case of the cure of epilepsy
by hypnotism appeared originally in "Observations on
Trance," and its later developments in the pamphlet on

" Magic and Witchcraft." It was regarded by Braid as a most noteworthy instance of the therapeutic value of his discovery.

"In July, 1849, I was called to attend Mrs ——, thirty years of age, married, and the mother of three children. She had suffered from attacks of epilepsy for four years, for which she had been under the care of several medical men. The attacks became a little less frequent for some time, but again increased in frequency and severity, until, at the period when I was consulted, notwithstanding she was under the constant care of a physician, and was taking medicine prescribed by him repeatedly every day, still she was getting worse, so that she had as many as twenty-eight fits daily, besides what occurred during the night, when her attendants were asleep. The day before I first operated on her she had thirteen violent fits in the space of eight hours ; and, when I first saw her, she was jaw-locked as a sequel to one of these epileptic attacks. Her friends told me, that whenever this state of locked-jaw occurred, they were never able to open her mouth for many hours, however much force they might apply to effect it. They were, therefore, not a little surprised to observe that, by one of my usual processes for reducing the cataleptic state of muscles during hypnotism or mesmerism, I was enabled, in *a few seconds*, to unlock her jaws and open her mouth, without the slightest difficulty or force. This patient was speedily thrown into the hypnotic condition by my usual method, and the result of my first operation, which was not more than of a quarter of an hour's duration, was this—that the patient had only *four* fits in eight hours, instead of the thirteen which she had during the corresponding period on the day previous.

"This patient was operated on by me twice each day subsequently, and on the fourth day the catamenia, which had been absent for several months, reappeared, and, in the course of a week, the fits were almost entirely arrested. Indeed, from that period, she had only three fits, and these were brought on by courses of considerable mental excitement. The catamenia recurred regularly for the next six months, and the results realised from the use of hypnotism, when they again became suspended, beautifully and conclusively illustrate the value of this mode of treatment in such affections.

"At six successive periods I had occasion to resort to hypnotism, for this purpose, and in *every* instance with *entire success*, by *a single operation of from ten to eleven minutes each.* Moreover, I ascertained that this important result could be effected without either touch or pass, but by mental concentration and direction, and management of the circulation alone. On two or three other occasions, when I hypnotised this patient for neuralgic or rheumatic pain in a leg or arm, through the mode of managing and directing the influence, the pains were immediately removed, without exciting any visible manifestation on *any* spinal organ or function."

Subsequently to the publication of this report, Braid resorted to the same process, in the same case, on six further occasions, with no variation in the success. "It now occurred to me," he says, "inasmuch as it was my opinion that the change in the physical action resulted *entirely* from the fixed mental attention of the patient, with a predominant idea and faith in the power of the processes, that it would be interesting to ascertain whether the result might not be realised by

mental concentration alone, when she remained wide
awake. On the 4th of April, 1851, I proposed to test
this. The requisite means were had recourse to for determining, with the utmost accuracy, the actual physical
condition of the patient *before* commencing the experiment. Four ladies and one gentleman were present.
Having requested the gentleman to note the time
accurately, the patient being seated in an easy-chair,
I addressed her in the following words, which were
heard by all present:—'Now, keep your mind firmly
fixed on what you know should happen.' All remained silent; and, in order to withdraw my own
mind as much as possible from the patient, I took
up a volume of 'Southey's Life,' and engaged myself
in reading it. At the end of eleven minutes I asked
her if the desired effect had been produced, to which
she replied she did not know; but, upon proper
examination, I had incontestable proof of the success
of my experiment. Next month she did not require a
repetition of the process; but, on the 2nd of June,
it was again tried, in the presence of two professional
gentlemen, and with equal success, as on the 4th of
April. On the 28th of July I again had occasion
to resort to the process with this patient; and I was
then *particularly* anxious that it should succeed, as a
lady was present who required similar aid, and a good
example in point was likely to assist me considerably
in influencing her the more certainly. My mind, therefore, was unusually intent on the accomplishment of
the desired result. At the expiration of eleven minutes
I inquired if it had taken effect, when she replied she
could not tell. On examination, however, it was ascertained that it had been a failure on this occasion.

The patient hereupon remarked that, before sitting down, she thought it would *not* ensue *that night.* I enquired, why? To which she replied, 'Because I could not *fix my mind* on it *to-night,* from having been put out of my way just before I came here.' To this I replied, 'Well, if you cannot fix your mind on the idea when awake, I know that I can command the requisite attention when you are asleep, and therefore I will hypnotise you.' This I did, at the same time exciting the circulation. It very soon became apparent, from the expression of her features and the movements of her body, that the spell was in active operation; and on arousing her in eleven minutes, there was positive proof adduced that the experiment had not been tried in vain ; and all went on satisfactorily subsequently.

"It merits special attention that, notwithstanding my mind was particularly active in *willing* the desired effect when the patient was awake, no success followed, because the requisite *mental* condition of the *subject* was wanting. This failure, therefore, was as positive proof in support of my theory as the successful results ; in fact, it was even more so. For, on other occasions, my mind had not been so intently desiring the immediate result, and now I was most actively engaged in willing it, and even with greater confidence, because of former successes. . . . Next month, but on the 8th of October, in the presence of her mother, the waking experiment was again tried with complete success. On the 8th of November, 1851, her mother and my esteemed friend Dr William Stevens, well known as the author of the Saline Treatment of Cholera and Yellow Fever, being present, I once more tried the experiment, with most complete success, in eight minutes. From that period it has never required to be repeated, as all has gone on in the natural course up to July, 1852. .... I may add that, on the 2nd of June, after exciting the function by mental concentration, whilst the patient was simply reposing in an easy-chair, by causing her to try it for two minutes more with the legs and arms extended, so as to quicken the circulation, a greatly increased effect was produced; proving the importance of regulating the circulation, for modifying the subsequent results."

APPENDIX III

Synopsis of Counter-Experiments undertaken by James Braid to illustrate his Criticism of Reichenbach.

HAVING proved to his own satisfaction that patients in the hypnotic state were not susceptible to a special influence emanating from magnets without suggestion on the part of the operator, Braid proceeded to experiments in the vigilant state. These were of two kinds : —(1) When the subjects had an opportunity of seeing what was being done, and expected something to happen ; (2) When they did not see, but supposed an operation was taking place, and consequently expected something.

General Results.

" With nearly all the patients I have tried, many of whom had never been hypnotised or mesmerised, when drawing the magnet or other object slowly from the wrist to the points of the fingers, various effects were realised, such as a change of temperature, tingling, creeping, pricking, spasmodic twitching, catalepsy of the fingers, or arm, or both ; and reversing the motion was generally followed by a change of symptoms, from the altered current of ideas thereby suggested. Moreover, if any idea of what might be expected existed in the mind previously, or was suggested orally, during the process, it was generally very speedily realised. The above patients being now requested to look aside, or a screen having been interposed, so as to prevent their seeing what was being done, and they being requested to describe their sensations during the repetition of the processes, similar phenomena were stated to be realised, even when there was nothing whatever done beyond watching them and noting their responses. They believed the processes were being repeated, and had their minds directed to the part, and thus the physical action was excited, so as actually to lead them to believe and describe their feelings as arising from external impressions."

Typical Cases.

I. " The above fact was most remarkably evinced in a young gentleman twenty-one years of age. I first operated in this manner on his right hand by drawing a powerful horse-shoe magnet over the hand, without contact, whilst the armature was attached. He immediately observed a sensation of cold follow the course of the magnet. I reversed the passes, and he felt it *less cold*, but he felt no attraction between his hand and the magnet. I then removed the cross-bar, and tried the effect with both poles alternately, but still there was no change in the effect, and decidedly no proof of attraction between his hand and the magnet. In the afternoon of the same day I desired him to look aside and hold his hat between his eyes and his hand, and observe the effects when I operated on him, whilst he could not see my proceedings. He very soon described a recurrence of the same sort of sensations as those he felt in the morning, but they speedily became more intense and extended up the arm, producing rigidity of the member. In the course of two minutes

z

354 BRAID ON HYPNOTISM

this feeling attacked the other arm, and to some extent the whole body, and he was, moreover, seized with a fit of involuntary laughter, like that of hysteria, which continued for several minutes—in fact, until I put an end to the experiment. His first remark was, 'Now this experiment clearly proves that there must be some intimate connection between mineral magnetism and mesmerism, for I was most strangely affected, and could not possibly resist laughing during the extraordinary sensations with which my whole body was seized, as you drew the magnet over my hand and arm.' I replied that I drew a very different conclusion from the experiments, as *I had never used the magnet at all*, nor held it, nor anything else, near to him; and that the whole proved the truth of my position as to the extraordinary power of the mind over the body, and how mental impressions could change physical action."

II. Another experiment was performed upon a gentleman, twenty-eight and a half years of age, in perfect health at the time. " I requested him to extend his right arm laterally, and let it rest on a chair with the palm upwards; to turn his head in the opposite direction, so that he might not see what I was doing; and to concentrate his attention on the feelings which might arise during my process. In about half-a-minute he felt an *aura* like a breath of air passing along the hand; in a little after, a slight pricking, and presently a feeling passed along the arm, as far as the elbow, which he described as similar to that of being slightly electrified. *All this, while I had been doing nothing*, beyond watching what might be realised. I then desired him to tell me what he felt NOW,—speaking in such a tone of voice as was calculated to lead him to believe I was operating in some different manner. The result was that the former sensations ceased ; but, when I requested him once more, to tell me what he felt *now*, the same sensations recurred. I then whispered to his wife, but in a tone sufficiently loud to be overheard by him, 'Observe now, and you will find his fingers begin to draw, and his hand will become clenched,—see how the little finger begins to move,'—and such was the case ; ' See the next one also going in like manner,'—and such effects followed ; and finally, the entire hand closed firmly, with a very unpleasant drawing motion of the whole flexor-muscles of the fore-arm. I did nothing whatever to this patient until the fingers were nearly closed, when I touched the palm of his hand with the point of my finger, which caused it to close more rapidly and firmly. After it had remained so for a short time, I blew upon the hand, which dissipated the previously existing mental impression, and instantly the hand became relaxed. The high respectability and intelligence of this gentleman rendered his testimony very valuable ; and especially so, when he was not only wide awake, but had never been either mesmerised, hypnotised, or so tested before." In another case, when the law had been explained to the subject and there was an attempted repetition of the experiment upon a different member, the effects took place less rapidly.

III. " A lady, thirty years of age, was requested to hold out her right hand over the arm of an easy-chair, whilst she turned her head to the left, to prevent her from seeing what I was doing, and to watch and describe to me the feelings she experienced in the hand during my process, which was to be performed without contact. She very soon felt a pricking in the point of

the third finger, which increased in intensity, and at
length extended up the arm. I then asked her how
her *thumb* felt, and presently the same feeling was
transferred to it ; and when asked to attend to the
middle of the fore-arm, in like manner the feeling was
presently perceived there. All the time I had been
doing *nothing* ; the whole was the result of her own
mind acting on her hand and arm. I now took the
large magnet, and allowed her to watch me drawing
it slowly over the hand, when the feeling was much as
before, only that she felt the cold from the steel when
brought very near to the skin. It was precisely the
same when closed as when opened, and the *same* sensations occurred when the north pole alone was approximated, or the south alone, or both together. She
experienced no sense of attraction between her hand
and the magnet from either pole, nor from both combined. I now requested this lady to keep a steady
gaze upon the poles of the large horse-shoe magnet,
and tell me if she saw anything (the room was not
darkened nor was the light strong), but nothing was
visible. I then told her to look steadily, and she would
see flame or fire come out of the poles. In a little
after this announcement she started, and said, 'Now I
see it, it is red ; how strange my eyes feel,' and instantly
she passed into the hypnotic state. This lady had
been repeatedly hypnotised. I now took the opportunity
of testing her as to the alleged power of the magnet
to attract her hand when asleep, but, as in the other
cases, the results were quite the contrary—the cold of
the magnet (and of either pole alike) caused her to
withdraw her hand the moment it touched her. I now
requested her to tell me what she *saw* (she being still

in the sleep). She said she still saw the red light. I desired her to put her finger to the place where she saw it. This she declined to do, being afraid that it would burn her. I thereupon assured her that it would not burn her, upon which she pointed *to the same place where the magnet was held before she went into the sleep*, instead of to where it was now held, which was near to her face, but towards the *opposite* side of the chair. This lady does not see from under her closed eyelids when hypnotised, as some patients do ; and the evidence her testimony affords in support of my opinion upon this subject is very conclusive, as she is a lady of very superior mental attainments, and one whose testimony merits unlimited confidence."

It will be observed that in the first case a sense-hallucination was produced by experimenting with the magnet, which was afterwards dispensed with in favour of simple suggestion ; in the second case the magnet was not used at all ; the third experiment began with simple suggestion, while the magnet was used subsequently. These selected instances exhaust therefore the possible variations in procedure. A cabinet experiment, which offers more close analogy with Reichenbach's method, may be cited in conclusion.

IV. " A lady, upwards of fifty-six years of age, in perfect health, and wide awake, having been taken into a dark closet, and desired to look at the powerful horseshoe magnet of nine elements, and describe what she saw, declared, after looking a considerable time, that she saw nothing. However, after I told her to look attentively, and she would see fire come out of it, she speedily saw sparks, and presently it seemed to her to burst forth, as she had witnessed an artificial represen-

tation of the volcano of Mount Vesuvius at some
public gardens. Without her knowledge, I closed
down the lid of the trunk which contained the magnet,
but still the same appearances were described as visible.
By putting leading questions, and asking her to describe what she saw from *another* part of the closet
(where there was nothing but bare walls), she went on
describing various shades of most brilliant coruscations
and flame, according to the leading questions I had put
for the purpose of changing the fundamental ideas.
On repeating the experiments, similar results were repeatedly realised by this patient. On taking her into
the closet after the magnet had been removed to another
part of the house, she still perceived the same visible
appearances of light and flame when there was nothing
but the bare walls to produce them ; and, two weeks after
the magnet was removed, when she went into the closet
by herself, the mere association of ideas was sufficient
to cause her to realise a visible representation of the
same light and flames. In like manner, when she was
made to touch the poles of the magnet when wide
awake, no manifestations of attraction took place between her hand and the magnet, but the moment the
idea was suggested that she would be held fast by its
powerful attraction, so that she would be utterly unable
to separate her hands from it, such result was realised ;
and, on separating it, by the suggestion of a new idea, and
causing her to touch the *other* pole in like manner, predi-
cating that *it* would *exert no attractive power* for the
fingers or hand, such negative effects were at once manifested. I know this lady was incapable of trying to deceive myself, or others present ; but she was self-deceived
and spell-bound by the predominance of a preconceived

idea, and was not less surprised at the varying powers of the instrument than others who witnessed the results."

By other experiments Braid found that a strong mental impression could produce the delusion of flame and light in broad daylight and during the waking state. He adds furthermore that " the same influence may be realised in respect to sound, smell, taste, heat, and cold," and all these not vaguely or generally, but after a vivid and discriminating manner.

V. The experiments of Reichenbach had been apparently transplanted to London, and when Braid paid a call on an " eminent and excellent physician," who used mesmerism occasionally in his practice, it was to find that he had experienced extraordinary effects from the use of magnets during that state. " He kindly offered to illustrate the fact on a patient who had been asleep all the time I was in the room, and in that stage during which I felt she could overhear every word of our conversation. He told me, that when he put the magnet into her hands, it would produce catalepsy of the hands and arms, and such was the result. He wafted the hands and the catalepsy ceased. He said that the mere touch of the magnet on a limb would stiffen it, and such he proved to be the fact. I now told him that I had a little instrument in my pocket, which, although far less than his, I felt assured would prove quite as powerful, and I offered to prove this by operating on the same patient, whom I had never seen before. . . . My instrument was about three inches long, the thickness of a quill, with a ring attached to the end of it. I told him that when put into her hands, he would find it catalepsize both hands and arms as his had done, and such was the result. Having reduced this by wafting, I took my instrument from her, and again returned it, *in another position*, and told him it would *now* have the very reverse effect—that she would not be able to hold it, and that although I closed her hands on it, they would open, and that it would drop out of them, and such was the case, to the great surprise of my worthy friend, who now desired to be informed *what I had done to the instrument to invest it with this new and apposite power.* This I declined doing for the present. . . . I now told him that a touch with it on either extremity would cause the extremity to rise and become cataleptic, and such was the result; that a second touch on the same joint would reduce the rigidity, and cause it to fall, and such again was proved to be the fact. After a variety of other experiments, every one of which proved precisely as I had predicted, she was aroused. I now applied the ring of my instrument to the third finger of the right hand, from which it was suspended, and told the doctor that when it was so suspended, it would send her to sleep. To this he replied, '*it never will*,' but I again told him that I felt confident. . . . We were then silent, and very speedily she was once more asleep. Having aroused her, I put the instrument on the second finger of her left hand, and told the doctor it would be found she could *not* go to sleep when it was placed there. He said he thought she would, and he sat steadily gazing at her, but I said firmly and confidently that she would not. After a considerable time the doctor asked her if she did not feel sleepy, to which she replied, 'not at all.' I then requested her to look at the point of the forefinger of her right hand, which I told the doctor would send her to sleep, and such was the result.

After being aroused, I desired her to keep a steady gaze at the nail of the thumb of the left hand, which would send her to sleep in like manner, and such proved to be the fact. Having repaired to another room, I explained to the doctor the real nature and powers of my little instrument—that it was nothing more than my *portmanteau-key and ring*, and that what had imparted to it such apparently varied powers was merely the prediction which the patient had overheard me make to him, acting upon her in the peculiar state of the nervous sleep, as irresistible impulses to be affected, according to the results she had heard me predict."

APPENDIX IV

A Posthumous Work of James Braid.

———

When Braid presented his published writings to the French Académie des Sciences, it will be seen from the introduction that they were accompanied by what appears to have been a summary of their contents in MS. Much about the same time another copy of this document was sent to Azam, which is recorded to have borne on the fly-leaf the following inscription :—" Presented to M. Azam as a mark of esteem and regard by James Braid, surgeon, Manchester, March 22, '60." It was therefore one of the last acts of the author, who died three days later. The MS. passed subsequently into the hands of a relative of Azam who in turn gave it to the American hypnotist, Dr G. M. Beard. Lastly, in August 1880, Beard confided it to W. Preyer, the German translator of Braid, and it was published in *Die Entdeckung des Hypnotismus*. An abstract of statements in the chief works of Braid, it is interesting only as the last literary production of the discoverer of hypnotism. It begins with an account of his observations on La Fontaine's experiments, undertaken as a sceptic with the view of discovering a deception. He next gives an account of the researches he undertook when he found that the phenomena were genuine and the conclusions to which they led him, together with the historical verification of his theory found subsequently in Ward's " History of the Hindoos." Some space is also devoted to his refutation of Reichenbach, to which experiments he seems always to have recurred with satisfaction. Among the practical points dwelt on are —(*a*) The absorption of the patient's mental powers by a fixed idea, to the exclusion of all influences which do not harmonise with the ruling conception ; (*b*) the connection between dream and hypnosis ; (*c*) the unsusceptibility of idiots, and of patients, otherwise sensitive, when under the influence of delirious fever ; (*d*) the important rôle played by imagination in hypnotic and kindred manifestations ; (*e*) the effect of magnetic passes ; (*f*) the entrancement of the mental powers in certain stages of the nervous sleep.

He recommends that the term Hypnotism should be applied only to those cases in which the patient on awaking remembers nothing that has occurred during the sleep, but can recall events more or less clearly when again entranced. Hypnotic coma should be used to designate that profounder state in which there is no recollection, the will-power suspended, and nothing remembered in a later sleep. Subjects who pass into this condition of coma are few in number, but at the same time they are the only individuals who can undergo painless surgical operations. In less intense stages there is consciousness of the operation, but not accompanied by violent pain. In such cases, however, the voice of a spectator, or a sudden draught may develop acute suffering.

Braid also prepared a collection of his most important researches, to be used as an appendix to a projected translation of " Neurypnology " by M. Masson, which design was afterwards abandoned.

APPENDIX V

A Short Bibliography of the Writings of James Braid.

I.

"Observations on Talipes, Strabismus, Stammering, and Spinal Contortion, and the best methods of Removing them." Published in the *Edinburgh Medical and Surgical Journal*, October 1841, vol. lvi. pp. 338-364. An illustrated account of a machine which Braid had invented for talipes on the principle of mechanical extension. He records that he had then operated on 246 club feet. He describes also his treatment of local paralysis by friction over the ganglia and gentle pressure over the course of the nerves leading to the paralysed parts. In the remarks on spinal contortion, it is stated that Braid was, with one exception, the first person to perform in England Guérin's treatment by division of the muscles of the back.

II.

Letter from James Braid, published in the *Medical Times*, March 12, 1842, giving some account of his lectures at the Hanover Square Rooms and the London Tavern. He reports the partial restoration to hearing of a deaf and dumb person, 32 years of age, and describes some experiments, with Mr Herbert Mayo, surgeon, of London, as his subject. The letter concludes with occasional remarks prompted by the criticisms and views of previous correspondents on the discovery of hypnotism. An interesting letter from Mr Mayo was afterwards published in the same periodical ; it was addressed, however, to Braid, and not to the editor. See " Biographical and Bibliographical Introduction " to this edition, p. 12.

III.

Letter from James Braid, published in the *Medical Times*, March 26, 1842, on the supposed power of seeing with other parts of the body than the eyes. It is also stated that patients are drawn or induced to obey the motions of the operator, not from any inherent magnetic power resident in the latter, but because their exalted state of feeling enables them to discern the currents of air set up by his gestures, and they advance and retire according to the direction of those currents. See " Introduction," p. 55.

IV.

" Satanic Agency and Mesmerism," a letter addressed to a clergyman, *i.e.*, the Rev. H. M'Neil, in reply to his strictures in a sermon. There is no trace of this pamphlet in the periodicals of the time, and it was, probably, unlike so many of Braid's minor publications, not a reprint from the columns of some medical periodical. It is mentioned by Mr C. W. Sutton, in the " Dictionary of National Biography," *s.v.* James Braid, and Preyer, the German translator of Braid's works, had evidently seen it, for he describes it as a fiery rejoinder. (*Die Entdeckung des Hypnotismus*, 1881, p. 1.) It is not, however, included in his

translation, and no copy is known to the present
writer. The sermon which gave occasion to the pamphlet was preached at S. Jude's Church, Liverpool, on
the evening of Sunday, April 10, 1842, and was published in the "Penny Pulpit," under the title of
"Satanic Agency and Mesmerism." It was noticed
in the *Phrenological Magazine*, Vol. xv. p. 288, the
reviewer observing that "the Rev. Hugh M'Neile
grants more power to the mesmerisers than they are
willing, it may be presumed, to accept the credit of, and
supposes them to have a co-operator not fit to name to
ears polite."

V.

A Letter entitled "Neuro-Hypnotism," published in
the *Medical Times*, July 9, 1842, concerning the refusal
of the Medical Section of the British Association to
hear his essay. See "Neurypnology," p. 84. With
some account of two cures, that of Miss Atkinson, *ibid.*,
281 *et seq.*, and another (unimportant).

VI.

"Neurypnology ; or, the Rationale of Nervous Sleep,
considered in Relation with Animal Magnetism. Illustrated by Numerous Cases of its successful application in the Relief and Cure of Disease." London :
John Churchill, Prince's Street, Soho. Adam and
Charles Black, Edinburgh, 1843, pp. xxii. and 266.
The motto on the title page is from Dugald Stewart:
"Unlimited scepticism is equally the child of imbecility
as implicit credulity." Compare Archbishop Whateley's "credulity of incredulity," the favourite maxim
of his friend Captain John James, the well-known non-
professional mesmerist. Dugald Stewart was Braid's

fellow-countryman, and seems to have been his chief philosophical guide, frequently quoted in his later pamphlets. "Neurypnology" was printed at Edinburgh by Andrew Shortrede, George IV. Bridge.

VII.

"Observations on the Phenomena of Phreno-Mesmerism," an article in the *Medical Times* of Nov. 11, 1843. It repeats the promise of further experiments on this subject made in the preface to "Neurypnology," and is prefaced by a letter, dated Nov. 4, which mentions the rapid sale of that work. See "Introduction," p. 16. This paper was reprinted in the *Phrenological Journal*, No. 78, Jan. 1874.

VIII.

A Letter of considerable length enclosing a contribution in continuation of the above subject. It was published separately in the *Medical Times* on Jan. 6, 1844. It discusses Elliotson's alleged mesmeric influence of metals, the Okey case (concerning the impugned *bona fides* of two of Elliotson's subjects), and the real nature of the verdicts pronounced by the various French commissions on Animal Magnetism.

IX.

"Observations on Some Mesmeric Phenomena," *Medical Times*, Jan. 13, 1844. The contribution which accompanied No. VII., and ostensibly in continuation of No. VI. The phenomena referred to are the exaltation of smell, hearing, and touch. The arguments are illustrated by cases. See "Introduction," p. 54.

X.

"Observations on Mesmeric and Hypnotic Phenomena." Two articles in the *Medical Times*, April 13 and 20, 1844. They are partly in reply to some questions proposed by a subscriber to that periodical, concerning the hypnotism of insane patients, clairvoyance, and the influence of contact. With the case of Mr Jones. See "Introduction," pp. 24-26. In the second paper Braid states that he had succeeded in teaching Greek, Latin, French, and Italian (? phrases) to subjects, which they remembered quite correctly when again hypnotised, but forgot in the waking state. There are also observations on the exaltation of sensibility to heat and cold.

XI.

Letter on the "Physiological Explanation of some Mesmeric Phenomena," *Medical Times*, August 31, 1844, being observations on an appreciative criticism of Hypnotism by James Singer, published in the Edinburgh Phrenological Journal. Chiefly in correction of errors.

XII.

A communication entitled "Remarks on Mr Simpson's Letter on Hypnotism," *Phrenological Journal*, October 1844. Mr Simpson's letter, which appeared in the issue for the previous July, was an able and extended account of Braid's theories by one who had made his acquaintance and witnessed some of his experiments. Braid's observations were designed to correct two errors, one as to the modes of exciting mental manifestations during the nervous sleep, the other as to the "reversed manifestations" in the third stage of hypnotism.

XIII.

" Experimental Inquiry to determine whether Hypnotic and Mesmeric Phenomena can be adduced in proof of Phrenology," *Medical Times*, November 30, 1844. These are the further experiments in PhrenoMagnetism promised in the preface to " Neurypnology " and in No. VI. An account of the results will be found in the " Introduction," pp. 19-29. This communication was reprinted in the *Phrenological Journal* for April 1845, with notes by the editor. In one of these it is said that the estimates of the strength of the manifestations, noted in accordance with the opinion of those present, had been sent to the writer, but that their publication did not seem to be necessary.

XIV.

" Case of Natural Somnambulism and Catalepsy, treated by Hypnotism, with Remarks on the Phenomena presented by Spontaneous Somnambulism, as well as that induced by various Artificial Processes." The only case of natural somnambulism which appears to have been treated by Braid, and it is so far interesting on that account. As an instance of the curative powers of hypnotism, it is not nearly so remarkable, however, as other cases treated by Braid, and is much too long to present. It occupied a considerable space in three issues of the *Medical Times* from October 26 to November 16, 1844, and was again recurred to in No. XIV. An apparent instance of clairvoyance in the subject, a female, is explained to the writer's satisfaction along his usual lines. The remarks on spontaneous somnambulism may be regarded as an extension of the views given at p. 123 of this edition of " Neurypnology."

XV.

"Magic, Mesmerism, Hypnotism, &c., Historically and Physiologically considered." A series of papers which appeared in consecutive issues of the *Medical Times* from December 7, 1844, until January 16, 1845. Portions were subsequently embodied in No. XXIX. For a description of these papers, see "Introduction," p. 30 *et seq.*

XVI.

"Case of Natural Somnambulism," as before, published in the *Medical Times* of May 17, 1845, being an account of further facts and inferences. Towards the close of the article the author promises a sequel to "Neurypnology" under the title of *Isis Interrogata*, containing a chapter devoted to the consideration of the extent to which mesmerism and hypnotism may be made available in mitigating or entirely preventing pain in important surgical operations. Also a full explanation of his theory of the phenomena of Phreno-Magnetism, namely, automatic muscular action excited through the reflex function of the nerves—suggesting ideas in the mind with which they are usually associated in the waking condition.

XVII.

Letter on the Fakeers of India, *Medical Times*, August 30, 1845, being an account of an inquiry, which the author was then undertaking, and submitting queries to elicit further information. These queries were also inserted in the *Lancet* and elsewhere. For the result of the Inquiry, see No. XXVI.

XVIII.

Dr Elliotson and Mr Braid, being three letters published by the latter in the *Medical Times* on Oct. 25, Nov. 1, and Nov. 8, 1845, in which he defends hypnotism from the accusation of coarseness. With a full account of the case of Miss Collins of Newark. See "Introduction," p. 31, and *Zoist*, 1845.

XIX.

"The Power of the Mind over the Body," *Medical Times*, June 13, 20, and 23, 1846. A full account of these important papers will be found in the "Introduction," and the substance of experiments in Appendix III.

XX.

"The Power of the Mind over the Body: an Experimental Inquiry into the Nature and Cause of the Phenomena attributed by Baron Reichenbach and others to a New Imponderable." London: John Churchill, Princess (*sic*) Street, Soho ; Adam & Charles Black, Edinburgh, 1846, pp. 36. Motto on title, "Truth is what it becomes us all to strive for." The pamphlet was printed by Grant & Co., Pall Mall, Market Street, Manchester. It is a reproduction, with very slight alterations, of the three papers published in the *Medical Times*.

XXI.

"Facts and Observations as to the Relative Value of Mesmeric and Hypnotic Coma, and Ethereal Narcotism, for the Mitigation or Entire Prevention of Pain during Surgical Operations," two papers published in the *Medical Times* on Feb. 13 and 27, 1847. A review of Esdaile's operations in India upon subjects in the mesmeric trance. It should be observed that hypnotism is here used for the first time by Braid as a synonym for mesmerism, but it is apparently by looseness of terminology, for he does not fail to state that he differs from his author entirely in the matter of theory. The superior convenience of ether over mesmeric and hypnotic methods is fully recognised, but there is otherwise nothing of importance in these contributions. These papers were reprinted by the *Edinburgh Medical and Surgical Journal*, April 1847.

XXII.

"Observations on the use of Ether for preventing pain during Surgical Operations, and the Moral Abuse which it is capable of being converted to," *Medical Times*, April 10, 1847. Occasioned by several fatal results which had recently followed the use of ether and were not referable to the operation itself. Including : (*a*) Cases of successful administration by Braid under difficult circumstances ; (*b*) Note on the frequency of spontaneous erotic manifestations during the primary stage of etherization ; (*c*) Absence of such manifestations during hypnotism.

XXIII.

" Mr Braid and Dr Elliotson," a letter in the *Medical Times* of Nov. 20, 1847, on some querulous paragraphs in the *Zoist*. Does not call for description. It is sufficient to say that the incident concerned exhibits the mesmeric leader in his usual unamiable light.

XXIV.

" Mr Braid and Mr Wakley," another letter in the *Medical Times*, Dec. 11, 1847, arising out of the above, and refuting a paltry accusation of plagiarism made against Braid by the editor of the *Lancet*. Braid has an easy task to show that it was ill-founded and ill-natured.

XXV.

"On the Use and Abuse of Anæsthetic Agents, and the best means of rousing patients who have been intensely affected by them," *Edinburgh Medical and Surgical Journal*, vol. 70, October 1848 (communicated by the author). The fatal accidents attending the use of chloroform are referred to carelessness occasioned by the uniform success in administration when the agent was first discovered. The paper is a sequel to No. XXII.

XXVI.

" Observations on Trance; or, Human Hybernation," three articles, published in the *Medical Times* on May 11, 18, and 25, 1850, and reprinted, with additions, as below.

XXVII.

"Observations on Trance ; or, Human Hyberna-tion." London : John Churchill, Princes Street, Soho ; Edinburgh: Adam & Charles Black, 1850, pp. vi. and 72. The mottoes on the title-page are *ex Oriente Lux*, and the following passage from Macnish :—" No affection, to which the animal frame is subject, is more remarkable than this (catalepsy or trance). There is such an apparent extinction of every faculty essential to life, that it is inconceivable how existence should go

on during the continuance of the fit." The printer was
William Tyler, Bolt Court, London. The work em-
bodies sections from No. XIII. For fuller account
see " Introduction," pp. 43-47.

XXVIII.

" Electro-Biological Phenomena Physiologically and
Psychologically considered." Originally delivered as
a lecture at the Royal Institution, Manchester, March
26, 1851, it appeared as an article in the *Monthly
Journal of Medical Science,* in June of that year, and
was finally printed in pamphlet form, of which no copy
is known, however, to the editor. For descriptive
account see " Introduction," pp. 47-52.

XXIX.

" Magic, Witchcraft, Animal Magnetism, Hypnotism,
and Electro-Biology." Third edition, greatly enlarged.
London : John Churchill ; Edinburgh : A. & C. Black,
1852, pp. 122. Printed by James T. Parkes, Cross
Street, Manchester. The motto on the title-page is
Amicus Plato, amicus Socrates, sed magis amica veritas.
The earlier editions of this work are unknown. See
" Introduction," pp. 52-55.

XXX.

" Entire Absence of Vagina, with Rudimentary State
of Uterus, and Remarkable Displacement of Rudimentary Ovaries and their Appendages, in a Married
Female, Seventy-Four Years of Age." Published in
the *Monthly Journal of Medical Science* for March 1853.
An account of an extraordinary case which came within
Braid's experience, but the paper contains naturally no
reference to hypnotism.

XXXI.

"Hypnotic Therapeutics, illustrated by cases," published in the *Monthly Journal of Medical Science* for July 1853, and immediately run off as a pamphlet without resetting. An appendix on "Table Turning" is added. There is no preface and no publisher. See "Introduction," pp. 56, 57.

XXXII.

Letter on "Table Turning," in illustration of the Muscular Theory, published anonymously in the *Manchester Examiner and Times*, April 30, 1853, and subsequently acknowledged by the writer. It adopts the same view as the appendix to "Hypnotic Therapeutics."

XXXIII.

"Electro-Biological Phenomena and the Physiology of Fascination." Including "The Critics Criticised." 1855. See "Biographical Introduction," p. 57.

XXXIV.

"On the Treatment of Certain Forms of Paralysis," 1855. *Ibid.*, p. 59.

N.B.—This Bibliography was in type before I had seen that of Dr J. Milne Bramwell (Proceedings of the Society for Psychical Research, Part 30, 1896), to which I am indebted for a few additional points.

INDEX

ABERCROMBIE, Dr, quoted, 159, 162, 163, 187, 215, 285, 286.
Academy of Medicine in France, 3.
Academy of Sciences, French, 61, 62, 362.
Alison, Professor, a supporter of Braid, 16.
Anæsthetics, analogy with Hypnotism, 74, 321.
Animal Magnetism, see Mesmerism.
Arago, his remarks on Mesmerism, 14.
Armstrong, Dr John, quoted, 154, 155.
Azam, B. F., experiments in Hypnotism, 61; work on, *ib.*; Braid's posthumous work presented to, 362.

BACCHANALIANS, whether under hypnotic influence, 132.
Beard, Dr G. M., American writer on Mesmerism, 362.
Bell, Sir Charles, 171.
Bennett, John Hughes, 16.
Berkeley, Bishop, 159.
Bertrand, his operations and views compared with those of Braid, 88, 90.
Binet and Féré on Animal Magnetism, v.
Braid, James, uneventful private life, 1; where born, *ib.*; to whom apprenticed, *ib.*; first practice, *ib.*; removal to Manchester, 2; date and cause of death, *ib.*; discovery nearly coincident with the report of the fourth French Commission on Mesmerism, 3; opportune nature of the discovery, 5; actual occasion of it, 6; Lafontaine's version of the matter, 7 *et seq.*; the same reviewed, 10, 11; Braid lectures in London, 11, 12; refutes a clergyman on the Satanic character of Mesmerism, 12; reception of his discovery, 12

et seq.; opposition of Elliotson, 16 *et seq.*; Braid's further observations on Phreno-Mesmerism, 19 *et seq.*; further curative experiments, 22; case of Mr Jones, 24, 26; Phreno-Magnetism again, 26 *et seq.*; Braid investigates the antiquities of Hypnotism, 30; his embroilment with the *Zoist*, 31; his counter-experiments on the New Imponderable of Reichenbach, 32 *et seq.* (see also Appendix III.); Braid collects evidence on Human Hibernation, 43 *et seq.*; cures rheumatism in his own person by Hypnotism, 45, 46; gives attention to Electro-Biology, 47 *et seq.*; replies to the strictures of Colquhoun, 52 *et seq.*; account of his last writings, 56 *et seq.*
British Association, rejects Braid's paper, 84; conversazione to members of, 83, 86.
Broca, M. P., experiments in Hypnotism, 61.
Brookes, II., lecturer on Animal Magnetism, quoted, 89, 90.
Brown, Dr Thomas, quoted, 158, 159.

CARPENTER, Dr, his opinion of Braid's discovery, 16; see also 32, 322.
Clarke, Dr Samuel Clarke, quoted, 164.
Coates, Professor, on Braid's personal influence, 18; on Phreno-Magnetism, 29; on lay professors of Mesmerism, &c., 326; on laboratory crimes, 325.
Colquhoun, J. C., 30, 52 *et seq.*, 324.
Courmelles, Dr Foveau de, 29, 325.
Cressfeld's " Value of Hypnotism " quoted, 18.
Criminal suggestion in Hypnotism, 325.

Cross-magnetising, 192, 193.
Cutaneous diseases cured by Hypnotism, 313, 314.

DARLING, Dr, propagator of Electro-Biology, 49.
Davy, Sir Humphrey, experiments with nitrous oxide, 73.
Deafness cured by Hypnotism, 239, 240.
Deaf and Dumb, Hypnotic experiments on, 240; Braid's views, 241-243; case of Nodan, 227, 244; case of John Wright, 244-247; case of James Shelmerdine, 247-254; case of Sarah Taylor, 254-257.
Deleuze, his critical history of Mesmerism, 3.
Distortions, Hypnotic treatment of, 306-308.
Disc, not used originally by Braid, 9.
Dreaming, compared with Somnambulism, 122, 123; causes of, 124, 125.
Duncan, a London lecturer on Braid's discovery, 11.
Durand de Gros, on the reception of Braid's discovery, 16; on its novelty, 19; on its connection with Electro-Biology, 48; his work on Braidism, 62, 63.

ELLIOTSON, Dr John, the central figure of Mesmerism in England, 4; his long silence concerning Braid's discovery, 18; his ungenerous attitude subsequently, 31, 32; on the influence of the will in Mesmerism, 167.
Electro-Biology, how far independent of Hypnotism, 6; its introduction from America, 47; its inventor, 6, 48; Braid's attitude regarding it, 48; his priority certain, ib.; Grimes acquainted with Hypnotism, 49; Braid's paper on the subject, 49-52; his later pamphlet, 57.
Engledere, Dr, on Phrenology, 20.
Ennemoser, Jos., his explanation of magical phenomena by Mesmerism, 31.
Epilepsy cured by Hypnotism, 305, 306, 346.

Eyes, effect of their consensual adjustment on pupils, 116.

FAKIRS, their entombment alive examined by Braid, 44; their performances explicable by Hypnotism, 103; Braid's inquiries, 370.
Faria, Abbé, mode of magnetising, 88.
Fascination, Braid's pamphlet on, 57; distinguished from catalepsy, ib.

GASPARIN, Count Agénor de, on the French Mesmeric Commission of 1784, 13.
Gregory, Dr W., his translation of Reichenbach, 33; his reply to Braid, 41; bears testimony to the results obtained by Braid, 16.
Grimes, J. S., the inventor of Electro-Biology, 6, 48, 49.

HALL, Dr Radclyffe, his reference to Braid in the Lancet, 15.
Hall, Spencer P., his lectures on Phreno-Magnetism, 174, 175; meeting with Braid, 182; witnesses Braid's phenomena, 183; his sub-divisions of the phrenological organs, 185; Braid explains his experiments, 186; examines his views, 190-193.
Headache cured by Hypnotism, cases of, 301-303, 347.
Heart, affections of, cured by Hypnotism, 308-310.
Heidenhain, 64.
Hibbert's "Philosophy of Apparitions" quoted, 74, 76.
Howitt, William, his account of Reichenbach, 32; on scientific attitude towards the Od force, 42, 43.
Hydrophobia, possible cure by means of Hypnotism, 81, 324.
Hypnotism separated from Animal Magnetism, 86 (compare 57, 58), 102; prejudices against, refuted, 91, 146; origin and definition of terms, 94, 327; modes of inducing, and success, 105-107, 109-111; causes of failure, 107; peculiar phenomena, 111, 130-133, 136-139; modes of de-

hypnotizing, 111, 128, 129, 329; hypnotism proceeds from a law of animal economy, 112, 113; rationale of, 113-126; wherein it differs from common sleep, 126-132; accompanied by no electric or magnetic change, 114; power of habit and imagination, 116, 117, 135; remarkable effects of a current of air, 117-120; influence or smell, 120; contrasted with ordinary reverie, 126, 127; probable cause of perfection of arts among the Greeks, 132; its phenomena consecutive, 72, 135, 136; its remarkable curative powers, 137-141 (also part ii. *passim*); its effects local or general at will, 144; its effects on mind as well as body, *ib.*; produces refreshing sleep, 147; sleep at will, 148; power of hearing faint sounds, 194; summary of Braid's position and views, 216, 217; certainty of the phenomena, 217; cause of, 219-221; degrees of susceptibility, 222, 223; how to employ hypnotism as a curative power, 225, 226.

IMAGINATION, effects of, 286, 338.
Insane patient, treatment of, 188, 189, 332.
Irritation, spinal, 303, 304.

KRAMER, his opinions criticised by Curtis, 245-247.

LAFONTAINE, visits to Manchester, 6-11; third visit, 12; see also 84, 98, 118, 121.
Lancet, The, on Mesmerism, 14, 15.
Laurent de Jussieu, 2.
Liebig's views as to the action of the alkaloids, 126.

MAGI as self-hypnotizers, 19; see also 30.
M'Nish on Reverie, quoted, 127, 133, 134.
Mayo, Dr Herbert, his testimony to Braid, 12; see also 106, 115, 155.
Materialism, arguments in refutation of, 151 *et seq.*; see p. 52.
Medical Times on Mesmerism, 15.

Mesmer, close of his career, 3; the French commission on his magnetic fluid, 13; experiments in Dr Franklin's garden, 102.
Mesmerism, periods of its history, 2, 3; condition in England at the time of Braid's discovery, 4, 5; the opposition which it has survived, 13; abuse of, 91; analogies with Hypnotism, 103; the higher phenomena, 103; rational and transcendental Mesmerism, 327.
Mind and Matter, how related to each other, 151.
Mind, the cause of organism, 160.
Müller, quoted, 125, 126, 130.
Music, its remarkable effects on Hypnotic patients, 131.

NEURALGIA of the heart cured by Hypnotism, 308, 309.

PALPITATION of heart cured by Hypnotism, 308, 309.
Paralysis cured by Hypnotism, cases of, 263-287, 344-346.
Passions excited by music, 164.
Phreno-Hypnotism,etc.; Phrenology not materialistic, 157; description of, *ib.*; what phrenology proves when illustrated by Hypnotism, 157; location of the organs, 166; excitation of the organs, 167; experiment, 168; size and position of organs, 169; inherent imperfection of phrenology, *ib.*; the reinforcement of Hypnotism, 170; objection considered, 171; localisation of function in the brain, 172; account of Braid's experiments, 174-210; mode of operating, 212.
Plato's opinion of mind and matter, 159.
Preyer, W., German translation of Braid's Works, v., vii., viii., 362.
Puséygur, Marquis de, his discovery of artificial somnambulism, 3.

QUINZATO, J. B., case cited, 331.

RHEUMATISM treated successfully by Hypnotism, 267-297, 343.
Reichenbach, Baron, account of his discovery, 32-34; Braid's criticism, 35; his counter-experiments, 36-43; synopsis of, 352-361.

Romanes, Professor, 64.
Royal Society of Medicine in France, 2.

SIGHT restored and improved by Hypnotism, 228-239.
Sleep, natural phenomena of, 122, 130, 131, 133, 134; causes of, 124.
Smell restored by Hypnotism, 258-260; case of olfactory hallucination, 340, 341; interesting experiment, 120.
Somnambulism, 123.
Spinal distortion cured by Hypnotism, 341-343.
Stewart, Dugald, quoted, 159.
Sunderland, Rev. le Roy, 190, 191.

Surgical operations during Hypnotism, 310.

TARANTULA Dance, 330.
Tetanus, 79-81, 314-316, 324.
Tic Doloureux cured by Hypnotism, 260-263.
Tonic Spasm cured by Hypnotism, 316, 317; see Tetanus.
Townshend, Chauncy Hare, 4, 31, 39.

VELPEAU, M., 61, 62.
Vitus's Dance, Saint, 330.

WHATELEY, Archbishop, an adherent of Mesmerism, 13; see also 366.

Zoist, The, 5, 18, 32.

TURNBULL AND SPEARS, PRINTERS, EDINBURGH.

www.ingramcontent.com/pod-product-compliance
Lightning Source LLC
Chambersburg PA
CBHW021354210326
41599CB00011B/877